MATHEMATICAL PARADISE

Paul Chika Emekwulu

Mathematical Paradise:
Getting to Know Triangular Numbers, Book One

Editing by Christine Rice Publishing Services

Cover Design by EJR Digital Art

Formatting by Wyrding Ways Press

Typeset by the author

Printed in the United States of America

Table of Contents

Preface

Conception and Approach of this Book

Over the years, I have listened to several inspirational and subliminal tapes. These implements and other personal development tools emphasize self-empowerment, possibility thinking, and other aspects of personal development. Among others, and in no particular order, I have read the following books and magazines on personal growth:

- *The Magic of Thinking Big* by David Joseph Schwartz

- *Psycho-Cybernetics* by Maxwell Maltz

- *The Tough-Minded Optimist* by Norman Vincent Peale

- *The Power of Your Subconscious Mind* by Joseph Murphy

- *Think and Grow Rich* by Napoleon Hill

- *How to Win Friends and Influence People* by Dale Carnegie

- *How to Stop Worrying and Start Living* by Dale Carnegie

- *Bring Out the Magic in Your Mind* by Al Koran

- *The Power of Positive Thinking* by Norman Vincent Peale

- *The Power of Miracle Metaphysics* by Robert B. Stone

- *The Magic of Believing* by Claude M. Bristol

- *Power through Constructive Thinking* by Emmet Fox

• *Move Ahead with Possibility Thinking* by Robert H. Schuller

• *Mind Power into the 21st Century: Techniques to Harness the*

Astounding Powers of Thought by John Kehoe

You may be wondering, "What is your point?" My point is that I have a penchant for exploring personal development topics as I do for mathematical concepts.

Let's talk about this book. To some people, everything about this book may be unusual or strange except maybe the title. You may also say that the content is unusual, because it is not written strictly in line with any traditional curriculum. You might also say the same about the chapter lengths, because of their briefness. You may also say that the general format is unusual.

You might be wondering whether this is a book on mathematics or a book on personal development. In both cases, you are right, because it is a book on both. Both topics are concerned with mind development. Generally, we can say that the whole book is strange. That is not an accident. I planned it that way.

Mathematical Paradise: Getting to Know Triangular Numbers was written on the premise of a famous American psychologist, Professor William James, who says that human beings use only 10 percent of their mental abilities. This raises the question: "Where is the rest?"

Answer:

- The rest is hidden.

- The rest is dormant.

The rest is, of course, limited by the five mundane or physical senses of sight, hearing, smell, touch, and taste. One of our weaknesses as human beings is our continued tendency to believe that physical reality can only be perceived through these five senses.

To some people, it is a matter of seeing it before believing. They want to hear it, smell it, touch it, and taste it. These people believe that any knowledge not gained through a conventional classroom should not only be questioned, but rejected outright and thrown out the window. This kind of thinking is not only dangerous, it is inimical to human progress and development.

In regards to the content in this book, before embarking on writing it, all I had was a formula for the nth triangular number, which is given by:

$$\frac{n(n+1)}{2}, \; n \geq 1.$$

I never had a course on triangular numbers in high school or college. In fact, such a course does not exist on either level.

Content Organization

Getting to Know Triangular Numbers, Book One is

divided into twelve short chapters.

Chapter One starts by defining triangular numbers as numbers of the form:

1, 3, 6, 10, 15, 21, 28, 36, 45, 55, 66, 78, 91, 105, 120, 136, 153, 171, 190, 210…

It also discusses the use of patterns in building a rule for the nth triangular number.

Additionally, it introduces us to the general form of expressing triangular numbers as:

$$\frac{n(n+1)}{2}, \; n \geq 1.$$

Representing triangular numbers using factorial notation is also a part of the investigation. Non-standard forms of representing the triangular numbers are also discussed. These non-standard forms represent odd- and even-subscripted triangular numbers. The chapter also explores a relationship between Fibonacci and triangular numbers.

Chapter Two is about deriving a formula for the number of diagonals in a given n-sided polygon. This formula is verified both algebraically and by use of examples. This chapter is also about a fundamental relationship between the number of diagonals in an n-sided polygon and first n triangular numbers.

Chapter Three revisits the number of diagonals in an n-sided polygon and first n triangular numbers. In this chapter, an expression for the number of sides of a polygon for a given number of diagonals is also explored. This is proved by using mathematical induction.

Chapter Four explores the transformation of non-standard forms into the standard form. This

chapter explores the fact that the two non-standard forms:

$2n^2-3n+1$, $n \geq 2$; and $n(2n+1)$, $n \geq 2$ represent triangular numbers of even subscripts.

Proving that both non-standard forms represent triangular numbers entail two levels of investigation:

Level 1and Level 2

Level 1: Translating $2n^2 - 3n+1$, $n \geq 2$ into the form $n(2n+1)$, $n \geq 1$

i.e. $2n^2-3n+1 \rightarrow n(2n+1)$

Level 2: Translating $n(2n+1)$ into the standard form:

$\dfrac{n(n+1)}{2}$, $n \geq 1$.

$n(2n+1) \rightarrow \dfrac{n(n+1)}{2}$

(Level 2)

$2n^2 - 3n+1 \rightarrow n(2n+1) \rightarrow \dfrac{n(n+1)}{2}$

Paul Emekwulu

(From Level 1 to Level 2 to standard form)

Both levels are investigated. Level 2 uses a form of logic called chain argument to conclude

its proof. It can also be shown that $2n^2 - 3n + 1$, $n \geq 2$ represents a triangular number by

showing that $2n^2 - 3n + 1$ can be transformed directly into $\frac{n(n+1)}{2}$,

thereby, omitting one of the levels of investigation.

This chapter also discusses and shows that $n(2n - 1)$, $n \geq 1$ represents odd-subscripted triangular numbers. These are triangular numbers of the form:

1, 6, 15, 28, 45, 66, 91, 120, 153, 190, 231, 276, 325, 378, 435, 496, 561 ...

There is only one level of investigation here:

$$n(2n-1) \rightarrow \frac{n(n+1)}{2} \qquad \text{(Level 1)}$$

Still this chapter discusses and shows that $2n^2+3n+1$, $n \geq 0$ represents triangular numbers of

odd subscripts. The goal is to show that such triangular numbers can be transformed into the

general form:

$$\frac{n(n+1)}{2},$$

$$\left[i.e.\ 2n^2 + 3n + 1 \to n(2n-1) \to \frac{n(n+1)}{2} \right].$$

(Level 1)

This chapter also investigates separately that:

$2n^2-3n + 1$, $n \geq 2$ and $n(2n-1)$, $n \geq 1$

represent triangular numbers of even and odd subscripts, respectively.

Chapter Five shows that $2n^2 - 3n + 1$ is a triangular number.

Chapter Six shows that $n(2n+1)$ is a triangular number.

Chapter Seven shows that $2n^2 - 3n + 1$ is a triangular number.

Chapter Eight discusses elementary proofs involving triangular numbers.

Chapter Nine proves that for any three consecutive Fibonacci numbers, a, b, c,

$$\frac{b^2 + 4ac - (c+a)}{2}$$

Paul Emekwulu

is a triangular number.

Chapter Ten proves that for any three consecutive triangular numbers,
$a, b, c, b(b-1) = ac.$

Chapter 11 uses basic algebraic principles to explore that the difference between the squares of two consecutive triangular numbers p and q with subscripts n and $n+1$, respectively, is equal to $(n+1)^3$.

Chapter Twelve proves that the sum of squares of any two consecutive triangular numbers a and b with subscripts n and $n+1$, respectively, is equal to a triangular number, a^2+b^2, whose subscript is equal to $(n+1)^2$ or $a + b$, or further still $(b-a)^2$.

Chapter Thirteen uses factorial notation of representing triangular numbers to prove that the sum of squares of any two consecutive triangular numbers a and b with subscripts n and $n+1$, respectively, is equal to a triangular number, a^2+b^2, whose subscript is equal to $(n+1)^2$ or $a + b$, or further still $(b-a)^2$.

In all the chapters, examples are given and the reader is encouraged to search for counter examples.

Following the thirteen chapters is the appendix, with combination tables of the first 46 triangular numbers, and Fibonacci and Lucas numbers. The list of seminars presented by the author himself and his internet presence are also featured there.

The subject of triangular numbers provided me an opportunity to not only challenge my mind and make a meaningful contribution to the mathematical community, but to also confirm the beauty, elegance, structure, and harmony inherent in mathematics.

June 2012
Paul Chika Emekwulu
Norman, OK
United States of America

What People are saying about this book

"I do know of books that deal with triangular numbers as a small chapter, unit, or lesson, mostly in a descriptive context, but I am not aware of any that delve into the patterns and relationships as deeply or thoroughly as your book does. I do confess that I have not truly researched the subject either. The mathematics is sound and well-developed, and the objectives listed for each article greatly support the topic of triangular numbers."

Penny Jackson
Lawton Public Schools
Secondary Education/Curriculum Specialist

"Father Ray Ackerman gave me two of your books, *Getting to Know Triangular numbers* and *Getting to Know Fibonacci numbers.* I found both of these books very interesting especially the one about Fibonacci numbers. If we ever offer a course in *Discrete Mathematics*, we would certainly consider your texts. In fact, a math teacher, graduate student or undergraduate student in an *Abstract Algebra* class would consider these texts as excellent reference books that have a lot of interesting materials in them that most students do not ever get to see." **Paul Buckelew, Oklahoma Center for Continuing Education, The University of Oklahoma**

CHAPTER 1

Recognizing Patterns

Objectives

At the end of the lesson, the students should be able to:

• Decompose a given numeral n into $n+1$ ways.

• Use partial sums of first n natural numbers to derive the rule for the nth triangular number.

• Find the nth triangular number given for any value of n.

• Derive $\dfrac{n(n+1)}{2}$ as the nth triangular number.

Introduction

What are triangular numbers and why are they called that?

Triangular numbers got their name, because they are represented by dots and arranged in a special format. They form equilateral triangles.

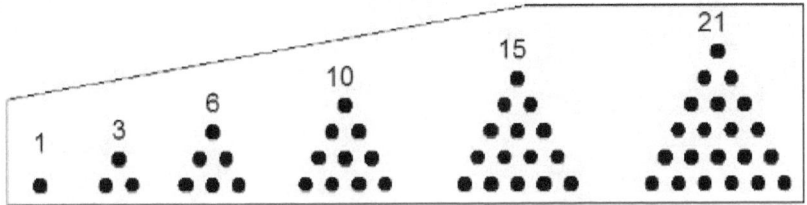

Figure 1: First Six Triangular Numbers Represented with Dots.

Standard Form of Representing Triangular

Numbers

Triangular numbers are generally written in the form:

$$\frac{n(n+1)}{2}$$

where $n = \{1, 2, 3, 4, 5, 6, 7, 8, 9, 10...\}$.

EXAMPLE 1:

When $n = 3$,

$$\frac{n(n+1)}{2} = \frac{3(3+1)}{2} = \frac{3(4)}{2} = 6.$$

EXAMPLE 2:

When $n = 5$,

$$\frac{n(n+1)}{2} = \frac{5(5+1)}{2} = \frac{5(6)}{2} = 15.$$

When $n = 5$, the fifth triangular number is 15.

EXAMPLE 3:

When $n = 6$,

$$\frac{n(n+1)}{2} = \frac{6(6+1)}{2} = \frac{6(7)}{2} = 21.$$

When $n = 6$, the sixth triangular number is 21.

EXAMPLE 4:

When $n = 7$,

$$\frac{n(n+1)}{2} = \frac{7(7+1)}{2} = \frac{7(8)}{2} = 28.$$

When $n = 7$, the seventh triangular number is 28.

EXAMPLE 5:

When $n = 8$,

$$\frac{n(n+1)}{2} = \frac{8(8+1)}{2} = \frac{8(9)}{2} = 36.$$

When $n = 8$, the eighth triangular number is 36.

EXAMPLE 6:

When $n = 9$,

$$\frac{n(n+1)}{2} = \frac{9(9+1)}{2} = \frac{9(10)}{2} = 45.$$

When $n = 9$, the ninth triangular number is 45.

Representing Triangular Numbers Using Factorial Notation

Using factorial notation, the nth triangular number can be written as:

$$\frac{(n+1)!}{2(n-1)!}. \text{ That is,}$$

$$\frac{(n+1)!}{2(n-1)!} = \frac{(n+1)(n)(n-1)!}{2(n-1)!} = \frac{n(n+1)}{2}.$$

EXAMPLE 1:

When $n = 3$,

$$\frac{(n+1)!}{2(n-1)!} = \frac{4!}{2(2)!} = \frac{4 \times 3 \times 2 \times 1}{2 \times 2 \times 1} = 6.$$

When $n = 3$, the third triangular number is 6.

EXAMPLE 2:

When $n = 5$,

$$\frac{(n+1)!}{2(n-1)!} = \frac{6!}{2(4)!} = \frac{6 \times 5 \times 4 \times 3 \times 2 \times 1}{2 \times 4 \times 3 \times 2 \times 1} = 15.$$

When $n = 5$, the fifth triangular number is 15.

EXAMPLE 3:

When $n = 6$,

$$\frac{(n+1)!}{2(n-1)!} = \frac{7!}{2(5)!} = \frac{7 \times 6 \times 5 \times 4 \times 3 \times 2 \times 1}{2 \times 5 \times 4 \times 3 \times 2 \times 1} = 21.$$

When $n = 6$, the sixth triangular number is 21.

EXAMPLE 4:

When $n = 7$,

$$\frac{(n+1)!}{2(n-1)!} = \frac{8!}{2(6)!} = \frac{8 \times 7 \times 6 \times 5 \times 4 \times 3 \times 2 \times 1}{2 \times 6 \times 5 \times 4 \times 3 \times 2 \times 1} = 28.$$

When $n = 7$, the seventh triangular number is 28.

EXAMPLE 5:

When $n = 8$,

$$\frac{(n+1)!}{2(n-1)!} = \frac{9!}{2(7)!} = \frac{9 \times 8 \times 7 \times 6 \times 5 \times 4 \times 3 \times 2 \times 1}{2 \times 7 \times 6 \times 5 \times 4 \times 3 \times 2 \times 1} = 36.$$

When $n = 8$, the eighth triangular number is 36.

EXAMPLE 6:

When $n = 9$,

$$\frac{(n+1)!}{2(n-1)!} = \frac{10!}{2(8)!} = \frac{10 \times 9 \times 8 \times 7 \times 6 \times 5 \times 4 \times 3 \times 2 \times 1}{2 \times 8 \times 7 \times 6 \times 5 \times 4 \times 3 \times 2 \times 1} = 45$$

When $n = 9$, the ninth triangular number is 45.

Decomposition of Numerals 1 to 9

Decomposition of numerals is another way we can derive the nth triangular number. If each of these numerals from 1 to 9 is represented by n, each numeral can be decomposed in $(n+1)$ ways. Of course, half the product of the number of ways of decomposing a numeral and the numeral itself is equal to a triangular number.

n	Number of Ways of Decomposing n	$n(n+1)$	$\dfrac{n(n+1)}{2}$
1	2	2	1
2	3	6	3
3	4	12	6
4	5	20	10
5	6	30	15

Table 1: Number of Ways of Decomposing n

1	2	3	4	5
0+1=1	0+2=2	0+3=3	0+4=4	0+5=5
1+0=1	2+0=2	3+0=3	4+0=4	5+0=5
	1+1=2	2+1=3	3+1=4	4+1=5
		1+2=3	1+3=4	1+4=5
			2+2=4	3+2=5
				2+3=5

Table 2: Decomposing the Numerals 1, 2, 3, 4, and 5

6	7	8	9	10
0+6=6	0+7=7	0+8=8	0+9=9	0+10=10
6+0=6	7+0=7	8+0=8	9+0=9	10+0=10
5+1=6	6+1=7	7+1=8	8+1=9	9+1=10
1+5=6	1+6=7	1+7=8	1+8=9	1+9=10
4+2=6	5+2=7	6+2=8	7+2=9	8+2=10
2+4=6	2+5=7	2+6=8	2+7=9	2+8=10
3+3=6	4+3=7	5+3=8	6+3=9	7+3=10
	3+4=7	3+5=8	3+6=9	3+7=10
		4+4=8	5+4=9	6+4=10
			4+5=9	4+6=10
				5+5=10

Table 3: Decomposing the Numerals 6, 7, 8, 9, and 10

Numeral	Sum of addends
1	1+0 = 1
2	2+0+1 = 3
3	3+0+2+1= 6
4	4+0+3+1+2 = 10
5	5+0+4+1+3+2 = 15
6	6+0+5+1+4+2+3 = 21
7	7+0+6+1+5+2+3+4 = 28

Table 4: Finding the nth Triangular Number by Decomposing Numerals

Table 4 reveals that the result of each addition fact is always a triangular number.

Numeral	Sum of addends
1	0+1 =1
2	0 + 2 + 1 = 3
3	0 + 3 + 1 + 2 = 6
4	0 + 4 + 1 + 3 + 2 = 10
5	0 + 5 + 1 + 4 + 2 + 3 = 15
6	0 + 6 + 1 + 5 + 2 + 4 + 3 = 21
7	0 + 7 + 1 + 6 + 2 + 5 + 3 + 4 = 28.

Table 5: Finding the nth Triangular Number by Decomposing Numerals

Table 5 reveals that the result of each addition fact is always a triangular number.

Partial Sums of First *n* Even Numbers

$0 + 2 = 2$

$0 + 2 + 4 = 6$

$0 + 2 + 4 + 6 = 12$

$0 + 2 + 4 + 6 + 8 = 20$

$0 + 2 + 4 + 6 + 8 + 10 = 30$

$0 + 2 + 4 + 6 + 8 + 10 + 12 = 42$

$0 + 2 + 4 + 6 + 8 + 10 + 12 + 14 = 56$

$0 + 2 + 4 + 6 + 8 + 10 + 12 + 14 + 16 = 72$

Dividing each of these partial sums by 2, results in a set of triangular numbers.

Take a look!

$$2 \div 2 = 1 \qquad 30 \div 2 = 15$$

$$6 \div 2 = 3 \qquad 42 \div 2 = 21$$

$$12 \div 2 = 6 \qquad 56 \div 2 = 28$$

$$20 \div 2 = 10 \qquad 72 \div 2 = 36$$

Proof that each S_n is a triangular number

Each partial sum can generally be proved to be a triangular number as follows:

Let each partial sum be represented by S_n.

$$S_n = \frac{n}{2}(a + L) \text{ where } a = 0, L = a + (n\text{-}1)d$$

By substitution,

$$S_n = \frac{n}{2}[a + a + (n-1)d]$$

$$= \frac{n}{2}[0 + 0 + (n-1)2]$$

$$= \frac{n}{2}(n-1)2 = n(n\text{-}1)$$

Since there is no restriction on n, if $S_n = n(n\text{-}1)$ is true for n, then it is true for $n+1$.

$$S_n = (n+1)[(n+1-1)]$$

$$= n(n+1)$$

When this expression is divided by 2 we have:

$$\frac{n(n+1)}{2}$$

Sum of Exponents of a and b Terms

Study the following expansions of $(a + b)^n$.

where $n = \{0, 1, 2, 3, 4, 5, 6, 7, 8, 9, 10...\}$.

Paul Emekwulu

$(a+b)^0 = 1$

$(a+b)^1 = a^1b^0 + a^0b^1$

$(a+b)^2 = a^2b^0 + 2a^1b^1 + a^0b^2$

$(a+b)^3 = a^3b^0 + 3a^2b^1 + 3a^1b^2 + a^0b^3$

$(a+b)^4 = a^4b^0 + 4a^3b^1 + 6a^2b^2 + 4a^1b^3 + a^0b^4$

$(a+b)^5 = a^5b^0 + 5a^4b^1 + 10a^3b^2 + 10a^2b^3 + 5a^1b^4 + a^0b^5$

$(a+b)^6 = a^6b^0 + 6a^5b^1 + 15a^4b^2 + 20a^3b^3 + 15a^2b^4 + 6a^1b^5 + a^0b^6$

$(a+b)^7 =$
$a^7b^0 + 7a^6b^1 + 21a^5b^2 + 35a^4b^3 + 35a^3b^4 + 21a^2b^5 + 7a^1b^6 + a^0b^7$

n	Sum of exponents of a terms
1	1+0 = 1
2	2+1+0 = 3
3	3+2+1+0 = 6
4	4+3+2+1+0 = 10
5	5+4+3+2+1+0 = 15
6	6+5+4+3+2+1+0 = 21
7	7+6+5+4+3+2+1+0 = 28
8	8+7+6+5+4+3+2+1+0 =36
9	9+8+7+6+5+4+3+2+1+0 = 45
10	10+9+8+7+6+5+4+3+2+1+0 = 55

Table 6: Sum of Exponents of a Terms

n	Sum of exponents of b terms
1	0+1=**1**
2	0+1+2=**3**
3	0+1+2+3= **6**
4	0+1+2+3+4=**10**
5	0+1+2+3+4+5=**15**
6	0+1+2+3+4+5+6 = **21**
7	0+1+2+3+4+5+6+7=**28**
8	0+1+2+3+4+5+6+7+8=**36**
9	0+1+2+3+4+5+6+7+8+9 = **45**
10	0+1+2+3+4+5+6+7+8+9+10 = **55**

Table 7: Sum of Exponents of b Terms

Relationship between Fibonacci Numbers and Triangular Numbers

For any three consecutive Fibonacci numbers a, b, and c,

$$\frac{b^2 + 4ac + (c + a)}{2}$$

is a triangular number, whose subscript is equal to $c+a$.

a	b	c	$b^2 + 4ac + (c + a)$	$\dfrac{b^2 + 4ac + (c + a)}{2}$	$c + a$ (Subscript)	Lucas Numbers
0	1	1	2	1	1	1
1	1	2	12	6	3	3
1	2	3	20	10	4	4
2	3	5	56	28	7	7
3	5	8	132	66	11	11
5	8	13	342	171	18	18
8	13	21	870	435	29	29

Table 8: Triangular Numbers with Subscripts as Lucas Numbers

The triangular numbers in Column 5 of Table 8 are those whose subscripts are the set of first n Lucas numbers. Lucas numbers are numbers of the form:

1, 3, 4, 7, 11, 18, 29, 47, 76, 123, 199, 322, 521, 843,

Paul Emekwulu

1364, 2207, 3571...

For any three consecutive Fibonacci numbers a, b, c,

$$\frac{b^2 + 4ac - (c+a)}{2}$$

is always a triangular number whose subscript is equal to $(c+a) - 1$.

Each resulting triangular number has a subscript of one less than a Lucas number, $c+a$.

a	b	c	$b^2+4ac-(c+a)$	$\dfrac{b^2 + 4ac - (c+a)}{2}$	$(c+a)-1$ (Subscript)	Lucas Numbers $(c+a)$
0	1	1	0	1	0	1
1	1	2	6	3	2	3
1	2	3	12	6	3	4
2	3	5	42	21	6	7
3	5	8	110	55	10	11
5	8	13	306	153	17	18
8	13	21	812	406	28	29

Table 9: Triangular Numbers with Subscripts of one Less than a Lucas Number

Finding the nth Triangular Number from Basic Principles

Let us create subscripts of the set of first n natural numbers from the above and designate them as A, B, C, D, E, and F.

$A = \{1\} = \{\Delta\}$

$B = \{1, 2\} = \{\Delta, \Delta\Delta\}$

$C = \{1, 2, 3\} = \{\Delta, \Delta\Delta, \Delta\Delta\Delta\}$

$D = \{1, 2, 3, 4\} = \{\Delta, \Delta\Delta, \Delta\Delta\Delta, \Delta\Delta\Delta\Delta\}$

$E = \{1, 2, 3, 4, 5\} = \{\Delta, \Delta\Delta, \Delta\Delta\Delta, \Delta\Delta\Delta\Delta, \Delta\Delta\Delta\Delta\Delta\}$

$F = \{1, 2, 3, 4, 5, 6\} = \{\Delta, \Delta\Delta, \Delta\Delta\Delta, \Delta\Delta\Delta\Delta, \Delta\Delta\Delta\Delta\Delta, \Delta\Delta\Delta\Delta\Delta\Delta\}$

We can represent the elements in each set by dots, which when arranged in a special format and joined will form equilateral triangles. Let us find the number of dots in each set. Counting the number of dots is the same as adding the natural numbers in each set. Each triangular number is a sum of a subset of the first n natural numbers. We can, therefore, represent each of the groups of dots as a partial sum of the set of the first n natural numbers. From that viewpoint, therefore:

$A = 1 = 1$

$B = 1 + 2 = 3$

$C = 1 + 2 + 3 = 6$

$D = 1 + 2 + 3 + 4 = 10$

$E = 1 + 2 + 3 + 4 + 5 = 15$

$F = 1 + 2 + 3 + 4 + 5 + 6 = 21$

For the numbers in set A, the sum is given by:

Paul Emekwulu

$$\frac{n}{2}(a+L).$$

Similarly, for the numbers in sets B, C, D, E, and F.

Therefore, by substitution:

$$\frac{n}{2}(a+L) = \frac{n}{2}\left[a+a+(n-1)d\right]$$

$$= \frac{n}{2}\left[2a+(n-1)d\right]$$

$$= \frac{n}{2}(2+n-1), \ a=1, \ d=1$$

$$= \frac{n}{2}(n+1)$$

Therefore, $\frac{n}{2}(a+L) = \frac{n}{2}(n+1)$.

Therefore, the nth triangular number is given by $\frac{n}{2}(n+1)$.

We can also arrive at the same conclusion by studying Tables 10 and 11.

n	a	L	$n+1$	$a+L$
1	1	1	1+1	1+1
2	1	2	2+1	1+2
3	1	3	3+1	1+3
4	1	4	4+1	1+4
5	1	5	5+1	1+5
6	1	6	6+1	1+6

Table 10: Showing that $n+1 = a+L$

Set	Number of Dots	a	n	L
A	$\frac{1}{2}(1+1)=1$	1	1	1
B	$\frac{2}{2}(1+2)=3$	1	2	2
C	$\frac{3}{2}(1+3)=6$	1	3	3
D	$\frac{4}{2}(1+4)=10$	1	4	4

Table 11: Showing that $x = y$,

where:

$$x = \frac{n}{2}(a + L) \text{ and } y = \frac{n(n+1)}{2}.$$

In Table 11, what property of addition can we use in order to enhance our effort in deriving the nth triangular number? Now, consider the set:

$$\left\{ \frac{1}{2}, \frac{2}{2}, \frac{3}{2}, \frac{4}{2}, \frac{5}{2}, \frac{6}{2} \right\}.$$

The numerators for the above fractions comprise the set of first n natural numbers.

Let this set be represented by n, while the set

$\{1 + 2, 1 + 3, 1 + 4, 1 + 5, 1 + 6...\}$

is represented by $n + 1$. (i.e. $n = \{1, 2, 3, 4, 5, 6...\}$)

and $n + 1 = \{1 + 2, 1 + 3, 1 + 4, 1 + 5, 1 + 6...\}$.

Therefore, the number of dots is given by the formula:

$$\frac{n}{2}(a+L).$$

By substitution, this is the nth triangular number.

So, the nth triangular number if given in terms of n is equal to:

$$\frac{n(n+1)}{2}.$$

A Search for Counter Examples

1. Can you identify any triangular number that cannot be represented in the standard form

    $$\frac{n(n+1)}{2}?$$

2. Can you identify any numeral that cannot be decomposed in $n+1$ ways?

CHAPTER 2

One-to-One Correspondence between the Number of Diagonals in an n-Sided Polygon and First n Triangular Numbers

Objectives

At the end of the lesson, the students should be able to:

- Discover a one-to-one correspondence between the number of diagonals in an n – sided polygon and first n triangular numbers.

- Derive a rule for finding the number of diagonals in an n-sided polygon.

Introduction

When we are required to find the number of diagonals in a simple n-sided polygon, (e.g. square, trapezium, etc.), what we can do is to draw a sketch of the given geometrical figure, join the vertices accordingly, and count the number of resulting diagonals.

Polygon	Number of Sides
Triangle	3
Square	4
Hexagon	6
Septagon	7

Table 12: List of Polygons and Corresponding number of

Paul Emekwulu

Diagonals

When n gets large, (e.g. when $n = 50$), drawing a sketch of the polygon, as well as the diagonals, becomes too cumbersome a task.

In this chapter, we shall investigate and explore a rule for finding the number of diagonals in a given n-sided polygon. With this rule, the task of finding the number of diagonals not only becomes easier, it becomes a part of us.

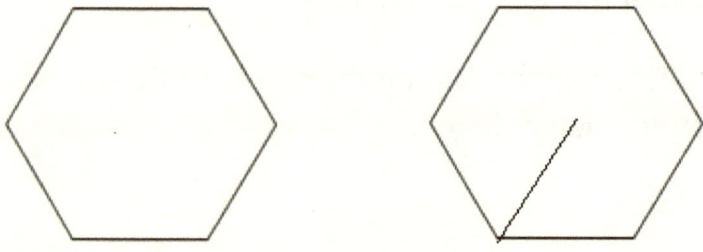

Figure 2: Drawing a Diagonal from one Vertex Only

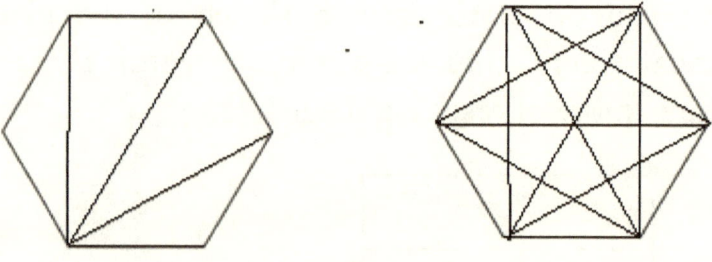

Figure 3: Drawing Diagonals from one and all Vertices

For each polygon, we shall do the following:

34

a) Find the number of diagonals from just one vertex. Call this n'.

b) Do the same for the rest of the vertices.

After step (b), the total number of diagonals becomes nn'.

n	n'	$n-n'$
3	0	3
4	1	3
5	2	3
6	3	3
7	4	3
8	5	3
9	6	3
10	7	3
11	8	3
12	9	3

Table 13: Finding the Relationship Between n and n'.

Justification of the Result

Relationship Between n and n'

n	$\dfrac{n(n-3)}{2}$	$\dfrac{(n-2)(n-1)}{2}$	n'
3	0	1	1
4	2	3	2
5	5	6	3
6	9	10	4
7	14	15	5
8	20	21	6
9	27	28	7
10	35	36	8

Table 14: Relationship between n and n'

From the above table:

$$n - n' = 2 \Leftrightarrow n' = n - 2.$$

By substitution in $\dfrac{n'(n'+1)}{2}$, we have:

$$\frac{n'(n'+1)}{2} = \frac{(n-2)[(n-2)+1]}{2}$$

$$= \frac{(n-2)(n-1)}{2}.$$

We can now set up an identity as follows:

In other words, if

$$\frac{n(n-3)}{2}$$

is true as the number of diagonals contained in an n-sided

36

polygon, and if

$$\frac{(n-2)(n-1)}{2}, n \geq 3$$

represents the n'^{th} triangular number, then

$$\frac{(n-2)(n-1)}{2} - \left\{\frac{n(n-3)}{2}\right\} \equiv 1.$$

(This is an identity, not an equation.)

Therefore, we have to simplify the LHS of the equation and see if it is equal to 1.

Doing so, we have:

$$\frac{(n-2)(n-1)}{2} - \left\{\frac{n(n-3)}{2}\right\} = \frac{n^2 - 3n + 2 - (n^2 - 3n)}{2}$$

$$= \frac{n^2 - 3n + 2 - n^2 + 3n}{2}$$

$$= \frac{(n^2 - n^2) - 3n + 3n + 2}{2} = 1.$$

Since

$$\frac{(n-2)(n-1)}{2} - \left\{\frac{n(n-3)}{2}\right\} = 1,$$

Paul Emekwulu

the following are true:

$$\frac{n(n-3)}{2}$$

is true as the number of diagonals in an n-sided polygon.

$$\frac{(n-2)(n-1)}{2}, n \geq 3$$

represents the n'^{th} triangular number.

The result has been verified.

For each n-sided polygon, there is a matching triangular number p and a corresponding number of diagonals q for which $p - q = 1$.

Another Form of Verification

Take a look at the following difference between successive number of diagonals in an n-sided polygon.

$2 - 0 = 2$	$20 - 14 = 6$
$5 - 2 = 3$	$27 - 20 = 7$
$9 - 5 = 4$	$35 - 27 = 8$
$14 - 9 = 5$	$44 - 35 = 9$

Since 2, 3, 4, 5, 6, 7, 8, 9…is a sequence,

0, 2, 5, 9, 14, 20, 27, 35…is likely to be a sequence

itself.

Let $\{3, 4, 5, 6, 7, 8, 9...\} = n'$ and also

let $\{2, 3, 4, 5, 6, 7, 8...\} = n$

Relationship between n and n'

$3 - 2 = 1$	$7 - 6 = 1$
$4 - 3 = 1$	$8 - 7 = 1$
$5 - 4 = 1$	$9 - 8 = 1$
$6 - 5 = 1$	$10 - 9 = 1$

These subtraction facts can be expressed as $n' - n = 1$.

Confirming the Above Relationship

If $n' - n = 1$, then $n' = n + 1$.

If $n' = n + 1$, then:

$$\frac{n'(n'-3)}{2} = \frac{(n+1)(n+1)-3}{2}$$

$$= \frac{(n+1)(n-2)}{2}$$

Let $\dfrac{n'(n'-3)}{2} = a_m$,

so that $a_{m+1} = \dfrac{(n'+1)(n'+1)-3}{2} = \dfrac{(n'+1)(n'-2)}{2}$.

Paul Emekwulu

(a_m and a_{m+1} are two consecutive terms of the sequence 0, 2, 5, 9, 14…).

By subtraction,

$$a_{m+1} - a_m = \frac{(n'+1)(n'-2)}{2} - \left[\frac{n'(n'-3)}{2}\right] = \frac{2n'-2}{2} = 2\left(\frac{n'-1}{2}\right) = n' - 1.$$

But since $n' = n + 1$, then $n' - 1 = (n + 1) - 1 = n$.

This confirms the fact that $n = \{2, 3, 4, 5, 6, 7…\}$

It also confirms the relationship between n and n', while

$$n = \frac{2n'-2}{2} = 2\left(\frac{n'-1}{2}\right) = n' - 1$$

But since $n' = n + 1$, $n' - 1 = (n+1) - 1 = n$.

EXAMPLE 1:

How many diagonals are there in a 10-sided polygon?

SOLUTION:

Number of diagonals is given by $\dfrac{n(n-3)}{2}$.

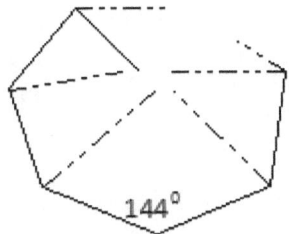

$$\frac{n(n-3)}{2} = \frac{10(10-3)}{2} = \frac{10\times7}{2} = 35.$$

EXAMPLE 2:

If a polygon has x diagonals, how many sides does the polygon have in terms of x?

SOLUTION:

Let the number of sides $= n$.

Therefore, $x = \dfrac{n(n-3)}{2}$.

By cross multiplication, we have:

$n^2 - 3n - 2x = 0$

For $n^2 - 3n - 2x = 0$, $a = 1$, $b = -3$, and $c = -2x$.

$$n = \frac{-b + \sqrt{9+8x}}{2}.$$

Therefore, by substitution, the number of sides is given by:

$$n = \frac{-(-3) + \sqrt{(-3)^2 - 4 \times 1 \times (-2x)}}{2 \times 1}$$

$$= \frac{3 + \sqrt{9 + 8x}}{2}$$

EXAMPLE 3:

The interior angle of a regular polygon is 120°. How many diagonals does the polygon have?

SOLUTION:

The sum of the interior angles of a polygon is given by (2*n*–4) right angles. Since the number of interior angles is equal to the number of sides, we have:

$$\left[\frac{2n - 4}{n}\right] 90° = 120$$

$$\Leftrightarrow 180n - 360 = 120n$$

$$\Leftrightarrow 60n = 360$$

$$\Leftrightarrow n = 6.$$

Number of diagonals = $\dfrac{n(n-3)}{2} = \dfrac{6(6-3)}{2} = 3 \times 3 = 9$

Answer: 9 diagonals.

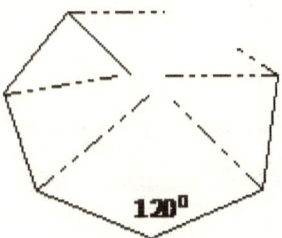

120°

Justification of Result

From Example 3, since $n = 6$, we can use our new formula to confirm the number of sides and, consequently, the number of diagonals.

Doing so, we have:

$$n = \frac{3 + \sqrt{9 + 8x}}{2}.$$

Therefore, substituting for $x = 9$, we have:

$$n = \frac{3 + \sqrt{9 + 8x}}{2} ?$$

$$= \frac{3 + \sqrt{9 + 72}}{2}, x = 9$$

$$= \frac{3 + \sqrt{81}}{2} = \frac{3 + 9}{2} = \frac{12}{2} = 6$$

A Search for Counter Examples

1. Can you identify any n-sided polygon whose number of diagonals cannot be found by using the formula:

Paul Emekwulu

$$\frac{n(n+1)}{2}?$$

2. Can you identify any n-sided polygon with a given x number of diagonals whose number of sides cannot be found from the formula?

$$n = \frac{3 + \sqrt{9 + 8x}}{2}?$$

3. Can you identify any n-sided polygon with x number of diagonals for which

$$n = \frac{3 + \sqrt{9 + 8x}}{2}?$$

is not a true statement?

Proof by Mathematical Induction

Before we start the proof, there is a need to find out how the number of diagonals relate to each other. Look at the following:

$$44 - 35 = 9 = 10 - 1 \qquad 14 - 9 = 5 = 6 - 1$$

$$35 - 27 = 8 = 9 - 1 \qquad 9 - 5 = 4 = 5 - 1$$

$$27 - 20 = 7 = 8 - 1 \qquad 5 - 2 = 3 = 4 - 1$$

$$20 - 14 = 6 = 7 - 1 \qquad 2 - 0 = 2 = 3 - 1$$

Table 15: Difference between Two Preceding Number of Diagonals

Also, look at the following:

$$2 = (3-1) + 0 \qquad 20 = (7-1) + 14$$

$$5 = (4-1) + 2 \qquad 27 = (8-1) + 20$$

$$9 = (5-1) + 5 \qquad 35 = (9-1) + 27$$

$$14 = (6-1) + 9 \qquad 44 = (10-1) + 35$$

How are these corresponding number of diagonals related to each other?

We will find out. But first, take a look at Table 15.

Table 15 shows the difference between two preceding number of diagonals.

We can see that each difference is equal to 1 less than n.

So, the equivalent expression for this difference in

terms of sides of polygon is $n - 1$.

Therefore, if the number of diagonals in an n-sided polygon is given by

$$\frac{n(n-3)}{2},$$

the next number of diagonals for $(n + 1)$-sided polygon is given by ·

$$\frac{n(n-3)}{2} + (n-1).$$

Therefore, the corresponding number of diagonals for 3, 4, 5, 6, 7, 8, 9, 10, 11...- sided

polygon can be listed generally as:

$$0, 2, 5, 9, 14, 20, 27, 35, 44... \quad \frac{n(n-3)}{2} \quad ... \quad \frac{n(n-3)}{2} + (n-1)$$

Now, we can resume the proof of the formula by mathematical induction.

Basic Step: Number of diagonals is given by:

$$\frac{n(n-3)}{2}.$$

$$P(n) = \frac{n(n-3)}{2}.$$

$P(3)$ is true since $\frac{3(3-3)}{2} = 0.$

Inductive Step:

$$\frac{n(n-3)}{2} + (n-1) = \frac{n(n-3) + 2(n-1)}{2}$$

$$= \frac{n^2 - 3n + 2n - 2}{2}$$

$$= \frac{n^2 - n - 2}{2}, \; n \geq 3$$

$$P(n+1) = \frac{(n+1)(n+1) - 3}{2}$$

$$= \frac{(n+1)(n-2)}{2}$$

$$= \frac{n^2 - 2n + n - 2}{2}$$

$$= \frac{n^2 - n - 2}{2}, \; n \geq 3.$$

$$\frac{n(n-3)}{2} + (n-1) = \frac{n(n-3) + 2(n-1)}{2}$$

$$= \frac{n^2 - 3n + 2n - 2}{2}$$

$$= \frac{n^2 - n - 2}{2}, \; n \geq 3$$

CHAPTER 3

Number of Diagonals in an *n*-Sided Polygon and Triangular Numbers

At the end of the lesson, the students should be able to:

- Derive a formula for finding the number of diagonals in an *n* sided polygon.

- Justify the formula for finding the number of diagonals in an *n*-sided polygon.

- Use examples to verify the formula for finding the number of diagonals in an *n*-sided polygon

Introduction

The sum of a subscript and a corresponding triangular number is equal to a member of the set

of numbers {2, 5, 9, 14, 20, 27...}. Call this *M*.

(i.e. subscript + the corresponding triangular number = *M*.)

$$n - 2 + \frac{n(n+1)}{2} = M$$

Number of Sides of Polygon (n)	$n-2$	$\dfrac{n(n+1)}{2}$	$n-2+\dfrac{n(n+1)}{2}$
3	3-2 = 1	1	2
4	4-2 = 2	3	5
5	5-2 = 3	6	9
6	6-2 = 4	10	14
7	7-2 = 5	15	20
8	8-2 = 6	21	27
9	9-2 = 7	28	35
10	10-2 = 8	36	44

Table 16

Wrong

Call the above Equation *(i)*.

What do you think about the new equation?

Does it truly represent the sum of a subscript and the corresponding triangular number?

If you thought, "I think so," you are wrong!

Here is why.

You are wrong because of the following reasons:

The quantity $n-2$ represents a subscript, while usually

$$\frac{n(n+1)}{2}$$

represents a triangular number.

But the "n" in $n - 2$ and the "n" in $\dfrac{n(n+1)}{2}$ are not the same variable.

Why? This is because the "n" in $n - 2$ refers to the set of numbers $\{3, 4, 5, 6, 7...\}$, while

the "n" in $\dfrac{n(n+1)}{2}$ refers to the set of numbers $\{1, 2, 3, 4, 5...\}$.

n	n'	$n' - n$
1	3	2
2	4	2
3	5	2
4	6	2
5	7	2
6	8	2
7	9	2
8	10	2

Table 17: Difference between n and n'

Differentiating Between the Variables

Since the two variables in $n - 2$ and $\dfrac{n(n+1)}{2}$ are not the same, let us denote the "n" in $\dfrac{n(n+1)}{2}$ as n', while the n in $n - 2$ remains unchanged.

So, $n = \{3, 4, 5, 6, 7, 8, 9...\}$ and $n' = \{1, 2, 3, 4, 5...\}$.

Now, let us go back and simplify Equation *(i)*.

Doing so, we have:

$$\frac{n-2}{1} + \frac{n(n+1)}{2} = \frac{2(n-2)+n(n+1)}{2}$$

$$= \frac{2n-4+n^2+n}{2}$$

$$= \frac{n^2+3n-4}{2} \quad \dots\dots\dots\dots\dots\dots\dots\dots\dots\dots\dots\text{(ii)}$$

Since the two variables, n and n', are not the same, what do you think about Equation *(ii)?*

If you thought, "It is wrong," you are correct.

What should we do?

This is where we refer back to the concept of n and n'.

We need a type of transformation.

Adding $n-2$ and $\frac{n(n+1)}{2}$ looks like adding apples and oranges.

- 5 apples + 5 oranges $\neq$ 10 apples or 10 oranges.

Paul Emekwulu

These are the various possibilities if and only if we can change an apple into an orange or an orange into an apple.

- 5 apples + 5 oranges = 10 apples
- 5 oranges + 5 oranges = 10 apples
- 5 apples + 5 oranges = 10 oranges
- 5 apples + 5 apples = 10 oranges

Unfortunately, we cannot transform an apple into an orange or an orange into an apple.

Therefore, the following is true:

- 5 apples + 5 apples = 10 apples
- 5 oranges + 5 oranges =10 oranges
- 5 apples + 5 oranges ≠ 10 apples
- 5 oranges + 5 oranges ≠ 10 apples
- 5 apples + 5 oranges ≠ 10 oranges
- 5 apples + 5 apples ≠ 10 oranges

An Advantage

The above equations do not add up, do they? The

answer is *no*. However, we still have an advantage here.

While we cannot transform an apple into an orange, and

an orange into an apple, we can transform n into n' or n'

into n by finding n in terms of n' or n' in terms of n.

How can this be done? Well, we can do it by

establishing a relationship between n and n'. While there

is a possible relationship between n and n', there is none

between an apple and an orange. This is obvious.

n	n'	$n-n'$
3	1	2
4	2	2
5	3	2
6	4	2
7	5	2
8	6	2
9	7	2

Table 18: Relationship between n and n'

From the above table, $n - n' = 2$ implies $n' = n - 2$.

In the context of this chapter, a general way of representing triangular numbers in terms of n

now is:

$$\frac{(n-2)[(n-2)+1]}{2} = \frac{(n-2)(n-1)}{2}$$

$$= \frac{n^2 - 3n + 2}{2}, \quad n \geq 3 \dots\dots\dots\dots\dots\dots\dots \ (iii)$$

By finding the relationship between n and n', we can express the number of diagonals in an n-

sided polygon in terms of n.

$n - n' = 2$ implies $n' = n - 2$

So, by substituting for n' in $n - 2 + \dfrac{n'(n'+1)}{2}$, we have:

$$= n - 2 + \frac{n'(n'+1)}{2} = n - 2 + \frac{(n-2)(n-2)+1}{2}$$

$$= \frac{2(n-2) + (n-2)(n-1)}{2}$$

$$= \frac{2(n-2) + n^2 - 3n + 2}{2}$$

$$= \frac{2n - 4 + n^2 - 3n + 2}{2}$$

$$= \frac{n^2 - n - 2}{2}$$

where $n = \{2, 3, 4, 5, 6, 7, 8, 9 \dots\}$

Therefore, the number of diagonals in an n-sided polygon is given by:

$$\frac{n'^2 - n' - 2}{2}, \ n' \geq 2.$$

Proof that the number of Diagonals is Given by:

$$\frac{n'^2 - n' - 2}{2}, \ n' \geq 2.$$

The proof can be done through the following steps:

Step 1: Tabulate the values of n and n' for which

$$\frac{n'^2 - n' - 2}{2} \ \text{and} \ \frac{n(n-3)}{2}$$

yield the same number of diagonals in an n-sided polygon.

Step 2: Isolate and tabulate the values of n and n'.

Step 3: Find n' in terms of n.

Step 4: Substitute for n' in $\frac{n'^2 - n' - 2}{2}$.

Step 5: Compare the result in step 4 with $\frac{n(n-3)}{2}$.

Step 6: If the result in step 4 is equal to $\frac{n(n-3)}{2}$, then $\frac{n'^2 - n' - 2}{2}$ represents the number of diagonals in an n-sided polygon where $n \geq 2$.

55

Now, let the investigation begin.

Step 1:

n'	$\dfrac{n'^2 - n' - 2}{2}$	n	$\dfrac{n(n+1)}{2}$
2	0	3	0
3	2	4	2
4	5	5	5
5	9	6	9
6	14	7	14
7	20	8	20
8	27	9	27

Table 19: Values of $\dfrac{n'^2 - n' - 2}{2}$ **and** $\dfrac{n(n+1)}{2}$

Step 2:

n'	n
2	3
3	4
4	5
5	6
6	7
7	8

Table 20

Step 3:

From Table 20, $n - n' = 1$.

From here, $n' = n - 1$.

Step 4:

By substituting for n' in $\dfrac{n'^2 - n' - 2}{2}$, we have:

$$\frac{n'^2 - n' - 2}{2} = \frac{(n-1)^2 - (n-1) - 2}{2}$$

$$= \frac{n^2 - 2n + 1 - n + 1 - 2}{2}$$

$$= \frac{n^2 - 3n + 1 + 1 - 2}{2}$$

$$= \frac{n^2 - 3n}{2} = \frac{n(n-3)}{2} , \quad n \geq 3$$

Step 5:

At this point, is $\dfrac{n'^2 - n' - 2}{2}$ equal to $\dfrac{n(n-3)}{2}$?

The answer is *yes*.

Step 6:

Since the answer to the question in step 5 is *yes*, then $\dfrac{n'^2 - n' - 2}{2}$ represents the number of diagonals in an n-sided polygon.

Showing that the nth Triangular Number is Given by $\dfrac{n'^2 - 3n' + 2}{2}, n \geq 2$

The proof can be done through the following steps:

Step 1: Tabulate the values of n and n' for which $\dfrac{n'^2 - 3n' + 2}{2}$ and $\dfrac{n(n+1)}{2}$ yield the same set of first n triangular numbers.

Step 2: Isolate and tabulate the values of n and n'.

Step 3: Find n' in terms of n.

Step 4: Substitute for n' in $\dfrac{n'^2 - 3n' - 2}{2}$.

Step 5: Compare the result in step 4 with $\dfrac{n(n+1)}{2}$.

Step 6: If the result in step 4 is equal to $\dfrac{n(n+1)}{2}$,

then $\dfrac{n'^2 - 3n' - 2}{2}$ represents the nth triangular number, where $n \geq 3$.

Now, let the investigation begin.

Step 1: Tabulate the values of n' and n for which $\dfrac{n'^2 - 3n' - 2}{2}$ and $\dfrac{n(n+1)}{2}$ yield the same set of first n triangular numbers.

n'	$\dfrac{n'^2 - 3n' - 2}{2}$	n	$\dfrac{n(n+1)}{2}$
3	1	1	1
4	3	2	3
5	6	3	6
6	10	4	10
7	15	5	15
8	21	6	21
9	28	7	28

Table 21: Values of $\dfrac{n'^2 - 3n' - 2}{2}$ and $\dfrac{n(n+1)}{2}$

Extracting values of *n* and *n*

n'	n	$n' - n$
3	1	2
4	2	2
5	3	2
6	4	2
7	5	2
8	6	2
9	7	2

Table 22: Relationship between *n* and *n'*

Step 3:

$n' - n = 2$.

From here, $n' = n + 2$.

Step 4:

By substituting for n' in $\dfrac{n'^2 - 3n' + 2}{2}$,

we have:

$$= \frac{(n+2)^2 - 3(n+2) + 2}{2}$$

$$= \frac{n^2 + 4n + 4 - 3n - 6 + 2}{2}$$

$$= \frac{n^2 + n + 4 - 6 + 2}{2}$$

$$= \frac{n^2 + n}{2} = \frac{n(n+1)}{2}$$

Step 5:

At this point, we have to ask ourselves this question: Is the result in step 4 equal to

$$\frac{n(n+1)}{2}?$$

The answer is *yes*.

Step 6 : Since the answer to the question in step 4 is *yes*, then

$$\frac{n'^2 - 3n' + 2}{2}$$

represents the *nth* triangular number where $n \geq 2$.

CHAPTER 4

Transforming Non-Standard Forms of Triangular Numbers into the Standard Form

Objectives

At the end of the lesson, the students should be able to:

• State the various forms of expressing triangular numbers in non-standard forms.

• Transform non-standard forms of nth triangular numbers to the standard form.

Introduction

Triangular numbers can be represented in a standard form as $\dfrac{n(n+1)}{2}$, regardless whether n is

even or odd.
For the purpose of this chapter, we shall, therefore,

regard $\dfrac{n(n+1)}{2}$ as the general standard form for

expressing the nth triangular number. There are other

forms of expressing the nth triangular number. Because

of their distinct nature, we shall regard these other

forms as non-standard forms. These non-standard forms may be classified into three:

Odd-Subscripted:
$n(2n-1), n \geq 1$

Even-Subscripted:
$n(2n+1), n \geq 1$

Both Odd and Even-Subscripted Terms Using Factorial Notation:
$\dfrac{(n+1)!}{2(n-1)!}, n \geq 1$

Those Representing Odd-Subscripted Triangular Numbers, $n(2n-1), n \geq 1$:

EXAMPLE 1:

When $n = 2$, $n(2n-1) = 2[(2 \times 2) - 1] = 2(3) = 6$.

EXAMPLE 2:

When $n = 4$, $n(2n-1) = 4[(2 \times 4) - 1] = 4(7) = 28$.

Those Representing Odd-Subscripted Triangular Numbers, $2n^2 + 3n + 1, n \geq 0$:

EXAMPLE 1:

When $n = 2$,

$$2n^2 + 3n + 1 = 2(2)^2 + 3(2) + 1$$
$$= 2(4) + 6 + 1 = 15.$$

EXAMPLE 2:

When $n = 4$,

$$2n^2 + 3n + 1 = 2(4)^2 + 3(4) + 1$$
$$= 2(16) + 12 + 1 = 45.$$

Those Representing Even-Subscripted Triangular Numbers, $2n^2 - 3n + 1$, $n \geq 2$:

EXAMPLE 1:

When $n = 2$,

$$2n^2 - 3n + 1 = 2(2)^2 - 3(2) + 1$$
$$= 2(4) - 6 + 1 = 3$$

EXAMPLE 2:

When $n = 4$,

$$2n^2 - 3n + 1 = 2(4)^2 - 3(4) + 1$$
$$= 2(16) - 12 + 1 = 21.$$

Those Representing Even-Subscripted Triangular Numbers, $n(2n + 1)$, $n \geq 1$:

EXAMPLE 1:

When $n = 2$,

$$n(2n + 1) = 2[(2 \times 2) + 1] = 2(5) = 10.$$

EXAMPLE 2:

When $n = 4$,

$$n(2n + 1) = 4[(2 \times 4) + 1] = 4(9) = 36.$$

We can show that these different forms represent triangular numbers by transforming them into the standard form. With truth-functional logic, we can show that each of the above non-standard forms, or a collection of them, represent triangular numbers by transforming each, or a collection, of these into the standard form. Assuming, for example, we are required to show that any of the following is true:

(a) $n(2n-1)$, $n \geq 1$ represent triangular numbers.

(b) $n(2n+1)$, $n \geq 1$ represent triangular numbers.

(c) $n(2n+1)$ and $2n^2 + 3n + 1$, $n \geq 0$ represent odd-subscripted triangular numbers.

(d) $n(2n-1)$, $n \geq 1$ and $2n^2 - 3n + 1$, $n \geq 2$ represent triangular numbers.

(e) $2n^2 + 3n + 1$, $n \geq 0$ represent odd-subscripted triangular numbers.

(f) $2n^2 - 3n + 1$, $n \geq 2$ represent even-subscripted triangular numbers.

Even if we succeed in showing the subscript of any of the above non-standard forms to be either even or odd, we still have the task of showing that non-standard forms are really another representation of triangular numbers. By this we mean that these different non-standard forms can be transformed into the general form.

To do this, these are the following steps:

Step 1: Note the standard form of a triangular number (T_n) as shown below:

$$T_n = \frac{n(n+1)}{2} \qquad \text{(i)}$$

Step 2: Differentiate Between the variables in either:

$n(2n - 1)$ or $n(2n + 1)$,

on one hand, and $\dfrac{n(n+1)}{2}$

on the other hand, as the case may be.

We can do this by letting the n in either

$n(2n - 1)$ or $n(2n+1)$ or $2n^2 - 3n + 1$

or $2n^2 + 3n + 1$ be n', while the n in

$$\frac{n(n+1)}{2}$$

remains unchanged.

Step 3: Tabulate the values of n' and n, for which $n(2n - 1)$

or $n(2n + 1)$

or $2n^2 - 3n + 1$

or $2n^2 + 3n + 1$

(again, as the case may be)

and $\dfrac{n(n+1)}{2}$

have different numerical results.

Step 4: Tabulate the values of n' and n, for which either

$n(2n - 1)$ or $n(2n + 1)$

or $2n^2 - 3n + 1$

or $2n^2 + 3n + 1$

and $\dfrac{n(n+1)}{2}$

have the same numerical results.

Step 5: Find the relationship between n and n'. This helps to establish an equation between n and n'. This relationship will help in completing the transformation in Step 6 (below) through substitution.

Step 6: Proceed with the required proof by substituting for n'.

Step 7: Simplify the expression in Step 6.

Step 8: Compare your result in Step 6 with the equation in Step 1. Call this Equation *(ii)*.

Step 9: If Equation *(ii)* is the same as Equation *(i)*,

$$\left[T_n = \frac{n(n+1)}{2} \right],$$

then the proof is complete, otherwise, go back to Step 8.

Step 10: Go to Step 1 and repeat Steps 3–8 until Equation *(ii)* is the same as:

$$\left[T_n = \frac{n(n+1)}{2} \right]$$

Let us look at some examples.

Showing that $2n^2 - 3n + 1$ and $n(2n+1)$ Represent Triangular Numbers

This involves two levels of investigation and a

conclusion. Using truth-functional logic, we can start and

complete the process. These two forms generate even-subscripted triangular numbers.

Level 1:

At this level, our goal is to show that triangular numbers of the form $2n^2 - 3n + 1$ can be transformed into triangular numbers of the form, $n(2n+1)$.

n	$2n^2 - 3n + 1$	n	$n(2n+1)$
1	0	1	3
2	3	2	10
3	10	3	21
4	21	4	36
5	36	5	55
6	55	6	78

Table 23: Different Numerical Results

n'	$2n'^2 - 3n' + 1$	n	$n(2n+1)$
2	3	1	3
3	10	2	10
4	21	3	21
5	36	4	36
6	55	5	55

Table 24: Same Numerical Results

At the second level, our goal is to show that $n(2n+1)$ can be transformed into the standard

form of $\dfrac{n(n+1)}{2}$.

Conclusion: If $2n^2-3n+1$ can be transformed into the form $n(2n+1)$, and $n(2n+1)$ can be

transformed into the form, $\dfrac{n(n+1)}{2}$, then $2n^2-3n+1$ has been transformed into the form,

$$\frac{n(n+1)}{2}.$$

This will then have completed the investigation.

The following propositions are relevant to our investigation.

If $2n^2-3n+1$ can be transformed into the form $n(2n+1)$ and $n(2n+1)$ can be transformed

into the form, $\frac{n(n+1)}{2}$, then $2n^2-3n+1$ has been transformed into the form, $\frac{n(n+1)}{2}$

If $n(2n+1)$ can be transformed into the form $\frac{n(n+1)}{2}$, and $2n^2-3n+1$ can be transformed into

the form, $n(2n+1)$, then $2n^2-3n+1$ has been transformed into the form, $\frac{n(n+1)}{2}$.

If $n(2n+1)$ can be transformed into the form $2n^2-3n+1$, and $2n^2-3n+1$ can be transformed into

the form, $\frac{n(n+1)}{2}$, then $n(2n+1)$ has been transformed into $\frac{n(n+1)}{2}$.

Now, let us expand on each of the above levels.

We will start with Level 1.

Level 1
Transforming $2n^2 - 3n + 1$ into the form $n(2n+1)$

(a) Tabulate values of n for which $2n^2 - 3n + 1$ and $n(2n+1)$ have different numerical results.

Why is this step necessary? It is necessary for it is an important step in finding a relationship between n and n'. The variables as seen in Table 23 are the same for $2n^2 - 3n + 1$ and $n(2n+1)$ even though the numerical results for $2n^2 - 3n + 1$ and $n(2n+1)$ are different.

(b) Tabulate values of n' and n for which $2n'^2 - 3n' + 1$ and $n(2n+1)$ have the same numerical results. The variables, as seen in Table 24, are not the same for $2n'^2 - 3n' + 1$ and $n(2n+1)$, but the numerical results for $2n'^2 - 3n' + 1$ and $n(2n+1)$ are the same.

n'	n	$n' - n$
2	1	1
3	2	1
4	3	1
5	4	1
6	5	1

Table 25: Relationship between n and n'

From the table, $n' - n = 1$ implies $n' = n + 1$.

Proceed with the required proof by substituting for n'.

Substituting for n' in $2n'^2 - 3n' + 1$, we have

71

Paul Emekwulu

$2n'^2 - 3n' + 1 = 2(n+1)^2 - 3(n+1) + 1$

$= 2(n^2 + 2n + 1) - 3(n+1) + 1$

$= 2n^2 + 4n + 2 - 3n{-}3 + 1$

$= 2n^2 + 4n - 3n + 2 - 3 + 1$

$= 2n^2 + n + (2{-}3{+}1)$

$= 2n^2 + n = n(2n{+}1)$

We have thus shown that $2n'^2 - 3n' + 1$ can be transformed into the non-standard form,

$n(2n{+}1)$. We can now proceed to the next level.

Level 2

Transforming $n(2n{+}1)$ into the form:

$$\frac{n(n+1)}{2}$$

The investigation is not over yet.

Level 2 starts where Level 1 ended.

At this point, we have to show that $n(2n{+}1)$ can be transformed into the standard form,

$$\frac{n(n+1)}{2}.$$

If we can show that $n(2n+1)$ represents a triangular number, by transforming it into the

standard form, $\dfrac{n(n+1)}{2}$, then by implication, $2n^2 - 3n + 1$ has also been transformed into the

general form, $\dfrac{n(n+1)}{2}$ and also, we have earlier shown that $n(2n + 1)$ and $2n'^2 - 3n' + 1$

represent triangular numbers.

This form of argument pattern is called "chain argument" in truth-functional logic.

If $A \rightarrow B$ and $B \rightarrow C$, then $A \rightarrow C$.

This can be arranged vertically as follows:

$A \rightarrow B$

and

$B \rightarrow C,$

therefore, $A \rightarrow C,$

where A = the non-standard form, $2n^2 - 3n + 1$ which can be transformed

into the form, $n(2n+1)$,

B = the non-standard form, $n(2n+1)$, which can be

transformed

into the general form,

$$\frac{n(n+1)}{2}.$$

and C = the non-standard form, $2n^2 - 3n + 1$, which can be transformed

into the general form:

$$\frac{n(n+1)}{2}.$$

Other argument patterns include modus tollens, modus

pollens, destructive dilemma, simplification, addition,

disjunctive argument, conjunction, and constructive

dilemma.

Remember, also, that in some way, as we have already seen, $n(2n+1)$ is related to

$2n'^2 - 3n' + 1$.

That makes sense. Now, let us continue.

Tabulate values of n for which $n(2n+1)$ and $\frac{n(n+1)}{2}$ have different numerical results.

Why is this step necessary?

It is necessary for three reasons:

(1) To show that the "n" in $n(2n+1)$ and the "n" in $\dfrac{n(n+1)}{2}$ are not the same variable.

(2) To show that, for the same values of n, $n(2n+1)$ and $\dfrac{n(n+1)}{2}$ do not have the same numerical results. This is as a consequence of (1) above.

Because of these different results, let the n in $n(2n+1)$ be n' and the n in $\dfrac{n(n+1)}{2}$ remains unchanged.

(3) To justify the need to differentiate between the variables in $n(2n+1)$ and $2n'^2 - 3n'+1$.

n	$n(2n+1)$	$\dfrac{n(n+1)}{2}$
1	3	1
2	10	3
3	21	6
4	36	10
5	55	15
6	78	21

Table 26: Values of n for which $n(2n+1)$ and $\dfrac{n(n+1)}{2}$ have Different Results

This takes us to the next step, which is tabulating values of n for which $n(2n + 1)$ and

$$\dfrac{n(n+1)}{2}$$

(unchanged) have the same numerical results. Why is this step an important part of the investigation? It is important, because it is a step towards finding a relationship between n and n'.

An important part of finding this relationship is isolating the values of n and n'.

n'	$n'(2n'+1)$	n	$\dfrac{n(n+1)}{2}$
1	3	2	3
2	10	4	10
3	21	6	21
4	36	8	36
5	55	10	55
6	78	12	78

Table 27: Values of n for which $n'(2n'+1)$ **and** $\dfrac{n(n+1)}{2}$ **have the Same Numerical Results**

When this is done, the next step is to establish an equation between n and n'.
Why do we need to find a relationship between n and n'?

Because it will help us in the transformation process, which consequently helps in concluding

our proof. Our goal is to show that the non-standard form $n(2n+1)$ can be transformed into

the standard form,

$$\frac{n(n+1)}{2}$$

Once we find this relationship, we are ready to substitute for n' in $n'(2n'+1)$.

Isolating n and n' and displaying them in a table, we have:

n'	n	$n - n'$
1	2	1
2	4	2
3	6	3
4	8	4

Table 28: Finding a Relationship between n and n'

From the above table, $n - n' = n'$, so that $2n' = n$ implies $n' = \dfrac{n}{2}$.

By substituting for n' in $n'(2n' + 1)$, we have:

$$n'(2n'+1) = \frac{n}{2}\left[2\left(\frac{n}{2}\right)+1\right]$$

$$= \frac{n}{2}[n+1] = \frac{n(n+1)}{2}.$$

Conclusion: Therefore, since $n(2n + 1)$ represents a triangular number, $2n^2 - 3n + 1$ also

represents triangular numbers of even subscripts. Since $n(2n+1)$ represents triangular

numbers of even subscripts and has been transformed into the general form, $\dfrac{n(n+1)}{2}$,

then $2n^2 - 3n + 1$ has also been transformed into the general form, and therefore, also

77

represents triangular numbers of even subscripts, since we have earlier shown that $n(2n +1)$

and $2n'^2 - 3n' + 1$ represent triangular numbers.

REMARK:

$2n'^2 - 3n' +1$ and $\dfrac{n(n+1)}{2}$

represent triangular numbers of different subscripts

while $2n'^2 - 3n' +1$ and $\dfrac{n(n+1)}{2}$

represent triangular numbers of the same subscripts.

EXAMPLE 1:

LEVEL 1

If $n = 4$, $n' = n + 1 = 4 +1 = 5$.

$2n'^2 - 3n' +1 = 2(5^2) - (3 \times 5) +1 = 50 - 15 + 1 = 36$.

LEVEL 2

If $n = 4$, $n(2n + 1) = 4[(2 \times 4) + 1] = 4 \times 9 = 36$.

Since the n in $n(2n + 1)$ and the n in $\dfrac{n(n+1)}{2}$ are not the same, let the n in $n(2n + 1)$

be n', while the n in $\dfrac{n(n+1)}{2}$ remains unchanged.

Conclusion: Therefore, if $n' = 4$, $n = 8$, and

78

$$\frac{n(n+1)}{2} = \frac{8 \times 9}{2} = 36.$$

The transformation is now complete, since at both Levels 1 and 2, the triangular numbers are the same.

EXAMPLE 2:
LEVEL 1

If $n = 6$, $n' = n + 1 = 6 + 1 = 7$.

$2n'^2 - 3n' + 1 = 2(7)^2 - (3 \times 7) + 1 = 98 - 21 + 1 = 78$

LEVEL 2

If $n = 6$, $n(2n + 1) = 6[(2 \times 6) + 1] = 6 \times 13 = 78$.

Since the n in $n(2n + 1)$ and the n in $\frac{n(n+1)}{2}$ are not the same, let the n in $n(2n + 1)$

be n', while the n in $\frac{n(n+1)}{2}$ remains unchanged.

Conclusion: Therefore, if $n' = 6$, $n = 12$, and

$$\frac{n(n+1)}{2} = \frac{12 \times 13}{2} = 78.$$

The transformation is now complete, since at both Levels 1 and 2, the triangular numbers are the same.

EXAMPLE 3:

LEVEL 1

If $n = 8$, $n' = n + 1 = 8 + 1 = 9$.

$2n'^2 - 3n' + 1 = 2(9^2) - (3 \times 9) + 1 = 162 - 27 + 1 = 136.$

LEVEL 2

If $n = 8$, $n(2n+1) = 8[(2 \times 8) + 1] = 136.$

Since the n in $n(2n+1)$ and the n in $\dfrac{n(n+1)}{2}$ are not the same, let the n in $n(2n+1)$ be n'

while n in $\dfrac{n(n+1)}{2}$ remains unchanged.

Conclusion: Therefore, if $n' = 8$, $n = 16$, and

$$\frac{n(n+1)}{2} = \frac{16 \times 17}{2} = 136$$

The transformation is now complete, since at both Levels 1 and 2, the triangular numbers are the same.

EXAMPLE 4:

LEVEL 1

If $n = 10$, $n' = n + 1 = 10 + 1 = 11.$

$2n'^2 - 3n' + 1 = 2(11^2) - (3 \times 11) + 1 = 242 - 33 + 1 = 210.$

LEVEL 2

If $n = 10$, $n(2n + 1) = 10[(2 \times 10) + 1] = 10 \times 21 = 210.$

Since the n in $n(2n + 1)$ and the n in $\dfrac{n(n+1)}{2}$ are not the same, let the n in $n(2n+1)$

be n' while the n in $\dfrac{n(n+1)}{2}$ remains unchanged.

Conclusion: Therefore, if $n' = 10$, $n = 20$, and

$$\frac{n(n+1)}{2} = \frac{20 \times 21}{2} = 210$$

The transformation is now complete, since at both Levels 1 and 2, the triangular numbers are the same.

EXAMPLE 5:

LEVEL 1

If $n = 12$, $n' = n + 1 = 12 + 1 = 13$.

$$2n'^2 - 3n' + 1 = 2(13^2) - (3 \times 13) + 1 = 338 - 39 + 1 = 300$$

LEVEL 2

If $n = 12$, $n(2n + 1) = 12[(2 \times 12) + 1] = 12 \times 25 = 300$.

Since the n in $n(2n + 1)$ and the n in $\dfrac{n(n+1)}{2}$ are not the same, let the n in $n(2n + 1)$

be n' while the n in $\dfrac{n(n+1)}{2}$ remains unchanged.

Conclusion: Therefore, if $n' = 12$, $n = 24$, and

$$\frac{n(n+1)}{2} = \frac{24 \times 25}{2} = 300.$$

Paul Emekwulu

The transformation is now complete, since at both Levels 1 and 2, the triangular numbers are the same.

A Search for Counter Examples

Can you identify an even-subscripted triangular number that can be expressed as

$2n^2 - 3n + 1$, which can be transformed into the non-standard form, $n(2n+1)$, but cannot be

transformed into the standard form, $\dfrac{n(n+1)}{2}$, or can be expressed as $2n^2 - 3n + 1$ but cannot be

transformed into the non-standard form, $n(2n+1)$, or into the standard form, $\dfrac{n(n+1)}{2}$?

CHAPTER 5

Show That $2n^2 - 3n + 1$ is a Triangular Number

Objectives

At the end of the lesson, the students should be able to:

- Show that $2n^2 - 3n + 1$ represents even-subscripted triangular numbers by transforming it

  into the form, $\dfrac{n(n+1)}{2}$.

- Use examples to show that $2n^2 - 3n + 1$ represents even-subscripted triangular numbers by

  transforming it into the form, $\dfrac{n(n+1)}{2}$.

Getting Started

If $2n^2 - 3n + 1$ is a triangular number, it can be transformed into the form, $\dfrac{n(n+1)}{2}$.

A Necessary Step

Let us start the investigation by tabulating values for n, for which $2n^2 - 3n + 1$ and

$$\frac{n(n+1)}{2}$$

will yield different numerical results.

n'	$2n'^2 - 3n' + 1$	n	$\dfrac{n(n+1)}{2}$
1	0	2	3
2	3	3	6
3	10	4	10
4	21	5	15
5	36	6	21

Table 29

This step is necessary because of two reasons:

(1) To show that the n in $\dfrac{n(n+1)}{2}$ and the n in $2n^2 - 3n + 1$ are not the same.

(2) To show that for the same value of n, $2n^2 - 3n + 1$ and $\dfrac{n(n+1)}{2}$ do not have the same

numerical results. This is as a consequence of (1) above.

Because of the different numerical results, let the n in $2n^2 - 3n + 1$ be n' while the n in

$$\frac{n(n+1)}{2}$$

remains unchanged. The next step is tabulating values of n and n', for which

$2n'^2 - 3n' + 1$ and $\dfrac{n(n+1)}{2}$ have the same numerical results.

Why is this step a necessary one?

It is necessary, because it is a step towards finding a relationship between n and n'.

How is n' related to n? Is there a relationship at all? The answer is *yes*.

Since the same value of n will yield different results for $2n^2 - 3n + 1$ and $\dfrac{n(n+1)}{2}$, we can rewrite $2n^2 - 3n + 1$ as $2n'^2 - 3n' + 1$.

Tabulating values of n' and n, for which $2n'^2 - 3n' + 1$ and $\dfrac{n(n+1)}{2}$ have the same numerical results we have:

n'	$2n'^2 - 3n' + 1$	n	$\dfrac{n(n+1)}{2}$
2	3	2	3
3	10	4	10
4	21	6	21
5	36	8	36
6	55	10	55

Table 30: Values of n and n' for which $2n'^2 - 3n' + 1$ and $\dfrac{n(n+1)}{2}$ have the same Numerical Results

Generally, the following is true:

n'	n	$n'-1$
2	2	1
3	4	2
4	6	3
5	8	4

Table 31: Relationship between n and n'

But specifically, the following is also true:

$$2(2-1) = 2 \quad 2(4-1) = 6$$

$$2(3-1) = 4 \quad 2(5-1) = 8$$

Generally, the above subtraction facts can be written as $2(n' - 1) = n$.

From here, $2(n'-1) = n \Leftrightarrow 2n'-2=n$ and

$$2n' = n+2 \Leftrightarrow n' = \frac{n+2}{2}$$

Substituting in $2n'-3n'+1$ for n', we have:

$$2n'^2 - 3n' + 1 = 2\left\{\frac{n+2}{2}\right\}^2 - 3\left\{\frac{n+2}{2}\right\} + 1$$

$$= 2\left\{\frac{n^2+4n+4}{4}\right\} - \left\{\frac{3n+6}{2}\right\} + 1$$

$$= \frac{n^2+4n+4-3n-6+2}{2}$$

$$= \frac{n^2+n}{2} = \frac{n(n+1)}{2}.$$

We have thus shown that $2n^2 - 3n + 1$ can be transformed into the standard form, $\dfrac{n(n+1)}{2}$. Therefore, the proof is complete.

A Search for Counter Examples

Can you identify an odd-subscripted triangular number that can be expressed in a non-standard form as $2n^2 - 3n + 1$ but cannot be transformed into the standard form, $\dfrac{n(n+1)}{2}$?

CHAPTER 6

Showing That $n(2n + 1)$ Represents a Triangular
Number

Objectives

- To show that $n(2n+1)$ represent even-subscripted
  triangular numbers.

- Use examples to show that $n(2n+1)$ represent
  odd-subscripted triangular numbers by

- transforming it into the form, $\dfrac{n(n+1)}{2}$.

Getting Started

- If $n(2n+1)$ represent a triangular number, it can be
  transformed into the general form,

$$\frac{n(n+1)}{2}.$$

There is only one level of investigation and a
conclusion here since we have to compare with

$$\frac{n(n+1)}{2}.$$

Different Numerical Results

We can start the investigation by tabulating values of n for which $n(2n+1)$ and $\frac{n(n+1)}{2}$ have different numerical results.

We have three reasons for doing this:

(1) To show that the "n" in $n(2n+1)$ and the "n" in $\frac{n(n+1)}{2}$ are different variables.

(2) To show that for the same values of n, $n(2n+1)$ and $\frac{n(n+1)}{2}$ do not yield the same

numerical results. This is as a consequence of (1) above.

(3) To justify the need to differentiate between the variable in $n(2n+1)$ and $\frac{n(n+1)}{2}$.

Tabulating values of n for which $n(2n+1)$ and $\frac{n(n+1)}{2}$ have different numerical values.

Doing so, we have:

Paul Emekwulu

n	$n(2n+1)$	$\dfrac{n(n+1)}{2}$
1	3	1
2	10	3
3	21	6
4	36	10
5	55	15
6	78	21

Table 32: Values of n for which n(2n+1) and $\dfrac{n(n+1)}{2}$ yield Different Results

Because of the difference in results, let the n in $n(2n +1)$ be n' while the n in $\dfrac{n(n+1)}{2}$ remains unchanged.

The Same Numerical Results
From here, the next step is tabulating values of n' and n for which $n'(2n' + 1)$ and $\dfrac{n(n+1)}{2}$ have

the same numerical results. Why is this step an important part of the investigation?

It is important because it is a step toward finding a relationship between n and n'.

Tabulate values of n' and n for which $n'(2n' + 1)$ and $\dfrac{n(n+1)}{2}$ have the same numerical results

Doing so, we have:

n'	$n'(2n'+1)$	n	$\dfrac{n(n+1)}{2}$
1	3	2	3
2	10	4	10
3	21	6	21
4	36	8	36

Table 33: Values of $n'(2n'+1)$ and $\dfrac{n(n+1)}{2}$ with Same Numerical Values

An important part of this relationship between n and n' is isolating the values of n and n'.

Having done this, the next step is to establish an equation between n and n'. This equation represents the relationship between n and n'.

Why do we need to find a relationship between n and n'? Because it is a necessary step in

concluding our proof. Our goal is to show that the non-standard form, $n(2n+1)$, can be

transformed into the standard form, $\dfrac{n(n+1)}{2}$.

Once we have found this relationship, we are ready to substitute for n' in $n'(2n'+1)$.

Isolating Values of n and n'

This isolation can be done in two ways.

Either way, the goal is to express n' in terms of n.

Paul Emekwulu

n'	n	$n - n'$
1	2	1
2	4	2
3	6	3
4	8	4
5	10	5
6	12	6

Table 34: Isolating Values of n and n'

From the above table, $n - n' = n'$, so that $2n' = n$, $n' = \dfrac{n}{2}$.

n'	n	$\dfrac{n}{2}$
1	2	1
2	4	2
3	6	3
4	8	4
5	10	5
6	12	6

Table 35

From the above table, $n' = \dfrac{n}{2}$.

By substituting for n' in $n'(2n' + 1)$, we have:

$$n'(2n' + 1) = \frac{n}{2}\left[2\left(\frac{n}{2}\right) + 1\right] = \frac{n}{2}[n+1] = \frac{n(n+1)}{2}.$$

92

CHAPTER 7

Showing that $2n^2 + 3n + 1$ Represents a Triangular Number

Objectives

- To show that $2n^2 + 3n + 1$ represent odd-subscripted triangular numbers by transforming it into the general form, $\dfrac{n(n+1)}{2}$.

- Use examples to show that $2n^2 + 3n + 1$, $n \geq 0$ represent odd-subscripted triangular numbers.

This involves one level of investigation and a conclusion. The goal is to show that triangular

numbers of the form $2n^2 + 3n + 1$ can be transformed into the standard form, $\dfrac{n(n+1)}{2}$.

Tabulate values of n for which $2n^2 + 3n + 1$ and $\dfrac{n(n+1)}{2}$ have different numerical results.

n	$2n^2+3n+1$	$\dfrac{n(n+1)}{2}$
0	1	0
1	6	1
2	15	3
3	28	6
4	45	10
5	66	15
6	91	21
7	120	28

Table 36: Different Numerical Results

A Necessary Step

Tabulate values of n for which $2n^2+3n+1$ and $\dfrac{n(n+1)}{2}$ have different numerical results.

Why is this step necessary?

It is necessary for three reasons:

(1) To show that the "n" in $2n^2+3n+1$ and the "n" in $\dfrac{n(n+1)}{2}$ are not the same variables.

(2) To show that for some values of n, $2n^2+3n+1$ and $\dfrac{n(n+1)}{2}$ do not have the same numerical results. This is as a consequence of (1) above.

(3) To justify the need to differentiate between the variables in $\dfrac{n(n+1)}{2}$ and $2n^2+3n+1$.

Because of these different results, let the "n" in $2n^2+3n$

+1 be n', while the "n" in

$$\frac{n(n+1)}{2}$$

remains unchanged.

This takes us to the next step of tabulating values of n for which $2n'^2 + 3n' + 1$ and

$$\frac{n(n+1)}{2}$$

have the same numerical results.

Tabulate values of n for which $2n'^2 + 3n' + 1$ and $\frac{n(n+1)}{2}$ have the same numerical results.

Why is this step an important part of the investigation?

It is important, because it is a step towards finding a relationship between n and n'.

n	$2n'^2 + 3n' + 1$	n	$\frac{n(n+1)}{2}$
1	6	3	6
2	15	5	15
3	28	7	28
4	45	9	45
5	66	11	66
6	91	13	91
7	120	15	120
8	153	17	153

Table 37: Same Numerical Results

An important part of finding this relationship is isolating the values of n and n'.

When this is done, the next step is to establish an equation between n and n'.

This equation represents the relationship between n and n'.

Why do we need to find a relationship between n and n'?

Because it is a necessary step in concluding our proof. Our goal is to show that the non-standard form, $2n^2 + 3n + 1$ can be transformed into the general form,

$$\frac{n(n+1)}{2}$$

Once we find this relationship, we are ready to substitute for n' in $2n'^2 + 3n' + 1$.

Isolating n and n' and displaying them in a table results in the following:

n'	n	$n - n'$
0	1	1
1	3	2
2	5	3
3	7	4
4	9	5
5	11	6
6	13	7
7	15	8
8	17	9

Table 38: Difference Between n and n'

From the above table, $n - n' = n' + 1$ implies $2n' = n - 1$.

From here,

$$n' = \frac{n-1}{2}.$$

Substituting for n' in $2n'^2 + 3n' + 1$, we have:

$$2n'^2 + 3n' + 1 = 2\left(\frac{n-1}{2}\right)^2 + 3\left(\frac{n-1}{2}\right) + 1$$

$$= 2\left(\frac{n^2 - 2n + 1}{4}\right) + 3\left(\frac{n-1}{2}\right) + 1$$

$$= \left(\frac{n^2 - 2n + 1}{2}\right) + 3\left(\frac{n-1}{2}\right) + 1$$

$$= \frac{n^2 - 2n + 1 + 3n - 3 + 2}{2}$$

$$= \frac{n^2 - 2n + 3n + 1 - 3 + 2}{2}$$

$$= \frac{n^2 + n}{2} = \frac{n(n+1)}{2}$$

We have thus shown that $2n^2 + 3n + 1$ can be transformed into the standard form:

$$\frac{n^2 + n}{2}.$$

Therefore, the proof is complete.

REMARK: We can start and complete this proof by transforming $2n^2 + 3n + 1$ into the non-

standard form, $n(2n - 1)$. Our hope is that when $n(2n-1)$ is transformed into the standard

form, then $2n^2 + 3n + 1$ has been transformed into the standard form,

$$\frac{n^2 + n}{2}.$$

Illustrating with Examples

EXAMPLE 1:

When $n = 11$

$$n' = \frac{11-1}{2} = 5.$$

$$2n'^2 + 3n' + 1 = 2(5)^2 + 3(5) + 1 = 66.$$

$$\frac{n(n+1)}{2} = 11\left(\frac{11+1}{2}\right) = 66.$$

EXAMPLE 2:

When $n = 13$

$$n' = \frac{13-1}{2} = 6.$$

$$2n'^2 + 3n' + 1 = 2(6)^2 + 3(6) + 1 = 91.$$

$$\frac{n(n+1)}{2} = 13\left(\frac{13+1}{2}\right) = 91.$$

A Search for Counter Examples

Can you identify an odd-subscripted triangular number that can be expressed in a non-

standard form as $2n^2 + 3n + 1$, but cannot be transformed into the standard form, $\frac{n(n+1)}{2}$?

CHAPTER 8

Prove that for any three consecutive Fibonacci numbers $a, b, c,$

$$\frac{b^2 + 4ac + (c+a)}{2}$$

is always a triangular number.

Objectives

At the end of the lesson, the students should be able to:

- Prove that for any three consecutive Fibonacci numbers $a, b, c,$

  $$\frac{b^2 + 4ac + (c+a)}{2}$$

  is always a triangular number.

- Use examples to show that for any three consecutive Fibonacci numbers $a, b, c,$

  $$\frac{b^2 + 4ac + (c+a)}{2}$$

  is always a triangular number.

Introduction

My deep interest and love for numbers of the Fibonacci

sequence were heightened by a dream I had on Tuesday, March 30, 1993. As I was studying and admiring the innocent – looking sequence and later on, triangular numbers, I started to notice some similarities in their properties which include the following:

- The sum of the squares of two consecutive Fibonacci numbers u_n and u_{n-1} is equal to a Fibonacci number, $(u_n)^2 + (u_{n-1})^2$ whose subscript is equal to $2n-1$.

In other words, $(u_n)^2 + (u_{n-1})^2 = u_{2n-1}$.

- The sum of the squares of two consecutive triangular numbers T_n and T_{n-1} is equal to a triangular number, $(T_n)^2 + (T_{n-1})^2$ whose subscript is equal to either of the following:

(a) $T_n + T_{n-1}$

(b) n^2

(c) $(T_n - T_{n-1})^2$

In other words, $(T_n)^2 + (T_{n-1})^2 = T_{T_n+T_{n-1}} = T_{n^2} = T_{(T_n - T_{n-1})^2}$.

These similarities in property led me to suspect a mathematical relationship between them. I was right. After a period of trial and error, I invented a mathematical formula that converts any three

101

consecutive Fibonacci numbers u_{n-1}, u_n, and u_{n+1} into a unique, single triangular number. This formula has two forms:

A separate formula for generating first *odd-subscripted triangular numbers*

These are triangular numbers of the form:

1, 6, 15, 28, 45, 66, 91, 120, 153...

A separate formula for generating first *even-subscripted triangular numbers*

These are triangular numbers of the form:

3, 10, 21, 36, 55, 78, 105, 136, 171,...

Despite what had already been said about Fibonacci numbers, people still at times, ask me whether the formulas apply to any three whole numbers. The answer is *no*. The three numbers must be *Fibonacci* numbers and they must be *consecutive*.

Proof

$$b^2 + 4ac = (c+a)^2$$

Therefore, by substitution for $b^2 + 4ac$ in $b^2 + 4ac + (c+a)$, we have:

$$\frac{b^2 + 4ac + (c+a)}{2} = \frac{(c+a)^2 + (c+a)}{2}$$

$$= \frac{(c+a)(c+a) + c + a}{2}$$

Let $(c + a) = n$.

By substitution for $(c + a)$ in the above, we have:

$$\frac{(c+a)(c+a) + c + a}{2} = \frac{(n)(n) + n}{2} = \frac{n(n) + n}{2}$$

$$= \frac{n(n+1)}{2}.$$

Illustrating with Examples

EXAMPLE 1
Consider 1, 2, and 3.

$$\frac{b^2 + 4ac + (c+a)}{2}$$

$$\frac{2^2 + (4 \times 1 \times 3) + (3+1)}{2}$$

$$= \frac{4 + 12 + 4}{2} = \frac{20}{2} = 10$$

103

EXAMPLE 2

Consider 2, 3, and 5.

$$\frac{b^2 + 4ac + (c + a)}{2}$$

$$\frac{3^2 + (4 \times 2 \times 5) + (5 + 2)}{2}$$

$$= \frac{9 + 40 + 7}{2} = \frac{56}{2} = 28$$

EXAMPLE 3

Consider 3, 5, and 8.

$$\frac{b^2 + 4ac + (c + a)}{2}$$

$$\frac{5^2 + (4 \times 3 \times 8) + (8 + 3)}{2}$$

$$= \frac{25 + 96 + 11}{2} = \frac{132}{2} = 66.$$

Examples with Two-Digit Fibonacci Numbers

EXAMPLE 1

Consider 13, 21, and 34.

$$\frac{b^2 + 4ac + (c+a)}{2}$$

$$\frac{21^2 + (4 \times 13 \times 34) + (34+13)}{2}$$

$$= \frac{441 + 1768 + 47}{2} = \frac{2256}{2} = 1128$$

EXAMPLE 2
Consider 21, 34, and 55.

$$\frac{b^2 + 4ac + (c+a)}{2}$$

$$\frac{34^2 + (4 \times 21 \times 55) + (55+21)}{2}$$

$$= \frac{1156 + 4620 + 76}{2} = \frac{5852}{2} = 2926$$

EXAMPLE 3
Consider 34, 55, and 89.

$$\frac{b^2 + 4ac + (c + a)}{2}$$

$$\frac{55^2 + (4 \times 34 \times 89) + (89 + 34)}{2}$$

$$= \frac{3025 + 12104 + 123}{2} = \frac{15252}{2} = 7626$$

A Search for Counter Examples

Can you identify any three consecutive triangular numbers a, b, c such that

$$\frac{b^2 + 4ac + (c + a)}{2}$$ is not always a Fibonacci number?

CHAPTER 9

Prove that for any three consecutive Fibonacci numbers

$a, b, c,$ $\quad \dfrac{b^2 + 4ac - (c+a)}{2}$

is always a triangular number.

Objectives

At the end of the lesson, the students should be able to:

• Prove that for any three consecutive Fibonacci numbers $a, b, c,$

$$\frac{b^2 + 4ac - (c+a)}{2}$$

is always a triangular number.

• Use examples to show that any three consecutive Fibonacci numbers $a, b, c,$

$$\frac{b^2 + 4ac - (c+a)}{2}$$

is always a triangular number.

Proof

$b^2 + 4ac = (c+a)^2$ for any three consecutive Fibonacci numbers $a, b, c.$

Therefore, by substituting in $\dfrac{b^2 + 4ac - (c + a)}{2}$ for $b^2 + 4ac$, we have:

$$\frac{b^2 + 4ac - (c + a)}{2} = \frac{(c + a)^2 - (c + a)}{2}$$

$$= \frac{(c + a)(c + a) - (c + a)}{2} \quad \cdots \cdots (i)$$

Let $c + a = n$.

By substituting for $c + a$ in Equation (i), we have:

$$\frac{b^2 + 4ac - (c + a)}{2} = \frac{(c + a)(c + a) - (c + a)}{2}$$

$$= \frac{(n)(n) - (n)}{2} = \frac{n^2 - n}{2}$$

$$= \frac{n(n - 1)}{2}.$$

By mathematical induction,

$$\frac{(n + 1)(n - 1) + 1}{2} = \frac{n(n + 1)}{2}$$

and $\dfrac{n(n - 1)}{2}$ is the nth triangular number.

Examples with One – Digit Fibonacci Numbers

EXAMPLE 1:

Consider 1, 2, and 3.

$$\frac{b^2 + 4ac - (c + a)}{2}$$

$$= \frac{2^2 + (4 \times 1 \times 3) - (3 + 1)}{2}$$

$$= \frac{4 + 12 - 4}{2}$$

$$= \frac{12}{2} = 6$$

EXAMPLE 2:

Consider 2, 3, and 5.

$$\frac{b^2 + 4ac + (c + a)}{2}$$

$$= \frac{3^2 + (4 \times 2 \times 5) - (5 + 2)}{2}$$

$$= \frac{9 + 40 - 7}{2}$$

$$= \frac{42}{2} = 21$$

Examples with Two-Digit Fibonacci Numbers

EXAMPLE 1	EXAMPLE 2
Consider 1, 2, and 3.	Consider 3, 5, and 8.
$$\frac{b^2 + 4ac - (c + a)}{2}$$	$$\frac{b^2 + 4ac + (c + a)}{2}$$
$$\frac{2^2 + (4 \times 1 \times 3) - (3 + 1)}{2}$$	$$\frac{5^2 + (4 \times 3 \times 8) - (8 + 3)}{2}$$
$$= \frac{4 + 12 - 4}{2}$$	$$= \frac{25 + 96 - 11}{2}$$
$$= \frac{12}{2}$$	$$= \frac{110}{2}$$
$= 6$	$= 55$

EXAMPLE 3

Consider 34, 55, and 89.

$$\frac{b^2 + 4ac + (c + a)}{2}$$

$$\frac{55^2 + (4 \times 34 \times 89) - (89 + 34)}{2}$$

$$= \frac{3025 + 12104 - 123}{2}$$

$$= \frac{15006}{2}$$

$$= 7503$$

Miscellaneous Example

Consider 55, 89, and 144.

$$\frac{b^2 + 4ac + (c + a)}{2}$$

$$= \frac{89^2 + (4 \times 55 \times 144) - (144 + 55)}{2}$$

$$= \frac{7921 + 31680 - 199}{2}$$

$$= \frac{39402}{2}$$

$$= 19701$$

A Search for Counter Examples

Can you identify any three consecutive Fibonacci numbers a, b, c such that

$$\frac{b^2 + 4ac - (c + a)}{2}$$

is not always a triangular number?

CHAPTER 10

For Any Three Consecutive Triangular Numbers a, b, c,
$b(b-1) = ac$

Objectives

At the end of the lesson, the students should be able to:

- Prove that for any three consecutive triangular numbers a, b, c, $b(b-1) = ac$.

- Use examples to show that for any three consecutive triangular numbers a, b, c, $b(b-1) = ac$.

- Search for counter examples of three consecutive triangular numbers a, b, c such that $b(b-1) = ac$ is not a true statement.

Proof

Let:

$$a = \frac{n(n+1)}{2}, b = \frac{(n+1)(n+2)}{2}, c = \frac{(n+2)(n+3)}{2}.$$

Let $ac = f(x)$.

Paul Emekwulu

By substitution we have:

$$b(b-1)=\left\{\frac{n^2+3n+2}{2}\right\}\left\{\frac{n^2+3n+2}{2}-1\right\}$$

$$=\left\{\frac{n^2+3n+2}{2}\right\}\left\{\frac{n^2+3n+2}{2}\right\}-\left\{\frac{n^2+3n+2}{2}\right\}$$

$$=\frac{n^2(n^2+3n+2)+3n(n^2+3n+2)+2(n^2+3n+2)}{4}-\left\{\frac{n^2+3n+2}{2}\right\}$$

$$=\frac{(n^4+3n^3+2n^2)+(3n^3+9n^2+6n)+(2n^2+6n+4)}{4}-\left\{\frac{n^2+3n+2}{2}\right\}$$

$$=\frac{n^4+6n^3+13n^2+12n+4}{4}-\left\{\frac{n^2+3n+2}{2}\right\}$$

$$=\frac{n^4+6n^3+13n^2+12n+4-2n^2-6n-4}{4}$$

$$=\frac{n^4+6n^3+13n^2+12n+4-2n^2-6n-4}{4}$$

$$=\frac{n^4+6n^3+13n^2-2n^2+12n-6n+4-4}{4}$$

$$=\frac{n^4+6n^3+11n^2+6n}{4}$$

Now let $\dfrac{n^4+6n^3+11n^2+6n}{4}=f(x)$.

Assuming $n+1$ is a factor of $f(x)$, then $f(-1) = 0$.

$$f(-1) = \frac{(-1)^4 + 6(-1)^3 + 11(-1)^2 + 6(-1)}{4}$$

$$= \frac{1 - 6 + 11 - 6}{4}$$

$$= \frac{-5 + 11 - 6}{4}$$

$$= \frac{6 - 6}{4} = 0$$

Therefore,

$$f(x) = \frac{n^4 + 6n^3 + 11n^2 + 6n}{4}$$

$$= \frac{(n+1)(n+2)(n^2 + 3n)}{4}$$

$$= \left[\frac{n(n+1)}{2}\right] \times \left[\frac{(n+2)(n+3)}{2}\right]$$

But $\dfrac{n(n+1)}{2} = a$ and $\dfrac{(n+2)(n+3)}{2} = c$.

Therefore $\left[\dfrac{n(n+1)}{2}\right] \times \left[\dfrac{(n+2)(n+3)}{2}\right] = ac$.

Therefore, $f(x) = ac$, and consequently, since $f(x) = b(b-1) = ac$, $b(b-1) = ac$.

Examples with One-Digit Triangular Numbers

EXAMPLE 1:

Consider 1, 3, and 6.

$b(b-1) = 3(3-1) = 6$.

$ac = 1 \times 6 = 6$.

EXAMPLE 2:

Consider 3, 6, and 10.

$b(b-1) = 6(6-1) = 30$.

$ac = 3 \times 10 = 30$.

Therefore, for any three consecutive triangular numbers a, b, c, $b(b-1) = ac$.

Examples with Two-Digit Triangular Numbers

EXAMPLE 1:

Consider 10, 15, and 21.

$b(b-1) = 15(15-1) = 210$.

$ac = 10 \times 21 = 210$.

EXAMPLE 2:

Consider 15, 21, and 28.

$b(b-1) = 21(21-1) = 420$.

$ac = 15 \times 28 = 420$.

Therefore, for any three consecutive triangular numbers a, b, c, $b(b-1) = ac$.

Examples with Three-Digit Triangular Numbers

EXAMPLE 1:

Consider 105, 120, and 136.

$b(b-1) = 120(120 - 1) = 14280$.

$ac = 105 \times 136 = 14280$.

EXAMPLE 2:

Consider 153, 171, and 190.

$b(b-1) = 171(171 - 1) = 29070$.

$ac = 153 \times 190 = 29070$.

Therefore, for any three consecutive triangular numbers a, b, c, $b(b-1) = ac$.

A Search for Counter Examples

Can you identify any three consecutive triangular numbers a, b, c such that $b(b-1)$ does not equal ac?

CHAPTER 11

Using Basic Algebraic Principles to Explore the Difference between the Squares of Two Consecutive Triangular Numbers

Objectives

At the end of the lesson, the students should be able to:

- Use algebra to prove that the difference between the squares of two consecutive triangular numbers, p and q with subscripts n and $n+1$, respectively, is equal to $(n+1)^3$

- Use examples to verify that the difference between the squares of two consecutive triangular numbers, p and q, with subscripts n and $n+1$, respectively, is equal to $(n+1)^3$

- Search for counter examples of two consecutive triangular numbers p, q with subscripts n and $n+1$ whose difference of squares is not equal to $(n+1)^3$.

Proof

Let the two consecutive triangular numbers be p and q.

Let $\dfrac{n(n+1)}{2} = p,$ and $\dfrac{(n+1)(n+2)}{2} = q$.

$$q^2 - p^2 = \left[\frac{(n^2 + 3n + 2)(n^2 + 3n + 2)}{4}\right] - \left[\frac{(n^2 + n)(n^2 + n)}{4}\right]$$

$$= \left[\frac{n^2(n^2 + 3n + 2) + 3n(n^2 + 3n + 2) + 2(n^2 + 3n + 2)}{4}\right] - \left[\frac{n^2(n^2 + n) + n(n^2 + n)}{4}\right]$$

$$= \left[\frac{n^4 + 3n^3 + 2n^2 + 3n^3 + 9n^2 + 6n + 2n^2 + 6n + 4}{4}\right] - \left[\frac{(n^4 + n^3) + (n^3 + n^2)}{4}\right]$$

$$= \left[\frac{n^4 + 6n^3 + 13n^2 + 12n + 4}{4}\right] - \left[\frac{n^4 + 2n^3 + n^2}{4}\right]$$

$$= \frac{6n^3 - 2n^3 + 13n^2 - n^2 + 12n + 4}{4}$$

$$= \frac{4n^3 + 12n^2 + 12n + 4}{4}$$

$$= \frac{4(n^3 + 3n^2 + 3n + 1)}{4}$$

$$= (n^2 + 2n + 1)(n + 1)$$

$$= (n + 1)(n + 1)(n + 1)$$

$$= (n + 1)^3$$

Paul Emekwulu

and $(n+1)$ is the subscript of q.

If the subscript of p is n, then the subscript of q is $(n+1)$.

Therefore, the difference between the squares of two consecutive triangular numbers, p and q with subscripts n and $n+1$, respectively, is equal to $(n+1)^3$.

Examples with One-Digit Triangular Numbers

EXAMPLE 1:

Consider 1 and 3, whose subscripts are 1 and 2, respectively.

Therefore, $n=1$ and $n+1=2$.

$3^2 - 1^2 = 9 - 1 = 8$ and $8 = (1+1)^3$.

EXAMPLE 2:

Consider 3 and 6, whose subscripts are 2 and 3, respectively.

Therefore, $n=2$ and $n+1=3$.

$6^2 - 3^2 = 36 - 9 = 27$ and $27 = (2+1)^3$.

Examples with Two-Digit Triangular Numbers

EXAMPLE 1:

Consider 36 and 45, whose subscripts are 8 and 9, respectively.

Therefore, $n=8$ and $n+1=9$.

$45^2 - 36^2 = 2025 - 1296 = 729$ and $729 = (8+1)^2$

EXAMPLE 2:

Consider 45 and 55, whose subscripts are 9 and 10, respectively.

Therefore, $n = 9$ and $n + 1 = 10$.

$55^2 - 45^2 = 3025 - 2025 = 1000$ and $1000 = (9+1)^2$

Examples with Three-Digit Triangular Numbers

EXAMPLE 1:

Consider 105 and 120, whose subscripts are 14 and 15, respectively.

Therefore, $n = 14$ and $n+1 = 15$.

$120^2 - 105^2 = 14,400 - 11025 = 3375$ and $3375 = (14+1)^2$

EXAMPLE 2:

Consider 136 and 153, whose subscripts are 14 and 15, respectively.

Therefore, $n = 16$ and $n+1 = 17$.

$153^2 - 136^2 = 23409 - 18496 = 4913$ and $4913 =$

$(16+1)^2$

A Search for Counter Examples

Can you identify any two consecutive triangular numbers p and q with subscripts n and $n+1$, respectively, whose difference of squares is not equal to $(n+1)^2$?

CHAPTER 12

The Sum of Squares of Any Two Consecutive
Triangular Numbers is a Triangular Number

Objectives

At the end of the lesson, the students should be able to:

- Prove that the sum of squares of any two
  consecutive triangular numbers a and b with
  subscripts n and $n + 1$, respectively, is equal to a
  triangular number, $a^2 + b^2$, where

$$a = \frac{n(n+1)}{2}, b = \frac{(n)(n+1)}{2}, a+b = \frac{(n+1)(n+2)}{2}.$$

- Use examples to verify that the sum of squares of
  any two consecutive triangular numbers a and b,
  whose subscripts n and $n + 1$, respectively, is
  equal to a triangular number, $a^2 + b^2$, whose
  subscript is equal to:

$$(n+1)^2 \text{ or } a+b \text{ or } (b-a)^2.$$

- Search for counter examples of consecutive
  triangular numbers a and b such that $a^2 + b^2$ is
  not a triangular number.

For any two consecutive Fibonacci numbers,

u_n and u_{n+1}, $(u_n)^2 + (u_{n+1})^2$ is a Fibonacci number, u_{2n+1} whose subscript is equal to the sum

of the subscripts of u_n and u_{n+1}.

Symbolically,

$$(u_n)^2 + (u_{n+1})^2 = u_{2n+1}.$$

This is familiar to a property also shared by triangular numbers, because the sum of squares of two consecutive triangular numbers, a and b, with corresponding subscripts, x and y, is also a triangular number, $a^2 + b^2$, whose subscript is equal to either $(a + b)$ or $(b–a)^2$ or $(y)^2$.

For Fibonacci numbers, the subscript $(u_n)^2 + (u_{n+1})^2$ is the sum of the subscripts of (u_n) and (u_{n+1}), or

$$(b-a)^2 = \left[\left\{\frac{(n+1)(n+2)}{2}\right\} - \left\{\frac{n(n+1)}{2}\right\}\right]^2$$

$$= \left\{\frac{n^2 + 3n + 2 - n^2 - n}{2}\right\}^2$$

$$= \left\{\frac{2n+2}{2}\right\}^2$$

$$= (n+1)^2.$$

Therefore, $(b-a)^2 = (n+1)^2$ and, consequently,

$$a+b = (b-a)^2 = b^2 - 2ab + a^2.$$

Verifying Our Result with Examples

Here are some specific examples designed to verify the above property.

Let the nth triangular number be equal to T_n.

Therefore, for any two consecutive triangular numbers, a and b, with subscripts

n and $n+1$, respectively, the sum of the squares of a and b is equal to a triangular number,

$a^2 + b^2$, whose subscript is equal to $(n+1)^2$ or $(a+b)$ or $(b-a)^2$.

Examples with One-Digit Triangular

Numbers

EXAMPLE 1:

If $a = 1$, $T_1 = 1$ and if $b = 3$, $T_2 = 3$.

$a^2 + b^2 = 1 + 9 = 10$

$a + b = 4$. But $10 = T_4$.

$(n+1)^2 = (1+1)^2 = 2^2 = 4$.

Or $(a+b) = 1 + 3 = 4$

Or $(b-a)^2 = (3-1)^2 = 2^2 = 4$

EXAMPLE 2:

If $a = 3$, $T_2 = 3$ and if $b = 6$, $T_3 = 6$.

$a^2 + b^2 = 9 + 36 = 45$

$a + b = 9$. But $45 = T_9$.

$(n+1)^2 = (2+1)^2 = 3^2 = 9$.

Or $(a+b) = 3 + 6 = 9$

Or $(b-a)^2 = (6-3)^2 = 3^2 = 9$

Examples with Two-Digit Triangular Numbers

EXAMPLE 1:

If $a = 15$, $T_5 = 15$ and if $b = 21$, $T_6 = 21$.

$a^2+b^2 = 225+441= 666$

$a+b =36$. But $666 = T_{36}$.

$(n+1)^2 = (5+1)^2 = 6^2 = 36$.

Or $(a+b)=15+21 = 36$

Or $(b-a)^2 = (21-15)^2 = 6^2 = 36$

EXAMPLE 2:

If $a =21$, $T_6 = 21$ and if $b = 28$, $T_7 = 28$.

$a^2+b^2 = 441+ 784 = 1225$

$a+b = 49$. But $1225 = T_{49}$.

$(n+1)^2 = (6+1)^2 =7^2 = 49$.

Or $(a+b)=21+28 = 49$

Or $(b-a)^2 = (28-21)^2 = 7^2 = 49$

Examples with Three-Digit Triangular Numbers

EXAMPLE 1:

If $a =105$, $T_{14} = 105$ and if $b =120$, $T_{15} = 120$.

$a^2+b^2 =11025+14400= 25425$

$a+b = 225$. But $25425 =T_{225}$.

$(n+1)^2 = (14+1)^2 =15^2 = 225$.

Or $(a+b) = 105+120 = 225$

Or $(b-a)^2 = (120-105)^2 = 15^2 = 225$

EXAMPLE 2:

If $a = 120$, $T_{15} = 120$ and if $b = 136$, $T_{16} = 136$.

$a^2+b^2 = 14400 + 18496 = 32896$

$a+b = 105+136 = 256$. But $32896 = T_{256}$.

$(n+1)^2 = (15+1)^2 = 16^2 = 256$.

Or $(a+b)^2 = 120+136 = 256$

Or $(b-a)^2 = (136-120)^2 = 16^2 = 256$

Therefore, for any two consecutive triangular numbers a and b whose subscripts are n and

$n + 1$, respectively, the sum of squares of a and b is equal to a triangular number, $a^2 + b^2$,

whose subscript is equal to $(n+1)^2$, or $a+b$ or $(b-a)^2$.

Alternative Proof Using Factorial Notation

Proof

If the first triangular number is $\dfrac{(n+1)!}{2(n-1)!}$,

then the next is:

$$\dfrac{(n+2)!}{2(n)!}.$$

Therefore, by addition, we have:

$$\dfrac{(n+1)!}{2(n-1)!} + \dfrac{(n+2)!}{2(n)!} = \dfrac{(n+1)(n)(n-1)!}{2(n-1)!} + \dfrac{(n+2)(n+1)(n)(n-1)!}{2(n)(n-1)!}$$

$$= \dfrac{n(n+1)}{2} + \dfrac{(n+1)(n+2)}{2}$$

$$= \dfrac{n(n+1)}{2} + \dfrac{(n+1)(n+2)}{2}$$

$$= \dfrac{n^2+n}{2} + \dfrac{n^2+3n+2}{2}$$

$$= \frac{n^2+n}{2} + \frac{n^2+3n+2}{2}$$

$$= \frac{2n^2+4n+2}{2}$$

$$= \frac{2(n^2+2n+1)}{2}$$

$$= n^2+2n+1 = (n+1)^2.$$

Since $\dfrac{(n+1)!}{2(n-1)!} + \dfrac{(n+2)!}{2(n)!} = (n+1)^2$,

the sum of two consecutive triangular numbers is equal to a square number, given that the nth triangular number is:

$$\frac{(n+1)!}{2(n-1)!}.$$

Illustrating with Examples

EXAMPLE 1

If $n = 4$, then $\dfrac{(n+1)!}{2(n-1)!} = \dfrac{(4+1)!}{2(4-1)!} = \dfrac{5 \times 4 \times 3 \times 2 \times 1!}{2 \times 3 \times 2 \times 1!} = 10$

and $\dfrac{(n+2)!}{2(n)!} = \dfrac{6!}{2(4)!} = \dfrac{6 \times 5 \times 4 \times 3 \times 2 \times 1!}{2 \times 4 \times 3 \times 2 \times 1!} = 15$

$10 + 15 = 25$ and 25 is a square number.

EXAMPLE 2

If $n = 5$, then $\dfrac{(n+1)!}{2(n-1)!} = \dfrac{(5+1)!}{2(5-1)!} = \dfrac{6 \times 5 \times 4 \times 3 \times 2 \times 1!}{2 \times 4 \times 3 \times 2 \times 1} = 15$

and $\dfrac{(n+2)!}{2(n)!} = \dfrac{7!}{2(5)!} = \dfrac{7 \times 6 \times 5 \times 4 \times 3 \times 2 \times 1!}{2 \times 5 \times 4 \times 3 \times 2 \times 1!} = 21$

15 + 21 = 36 and 36 is a square number.

Therefore, the sum of any two consecutive triangular numbers is equal to a square number,

given that $\dfrac{(n+1)!}{2(n-1)!}$ is a triangular number.

A Search for Counter Examples

Can you think of any two consecutive triangular numbers, a and b, with subscripts n and $n + 1$, respectively, whose sum of squares is not equal to a triangular number, $a^2 + b^2$, whose subscript is equal to $(n+1)^2$ or $(a+b)$ or $(b-a)^2$?

Appendix A

n	nth Triangular Number	nth Fibonacci Number	nth Lucas Number
1	1	1	1
2	3	1	3
3	6	2	4
4	10	3	7
5	15	5	11
6	21	8	18
7	28	13	29
8	36	21	47
9	45	34	76
10	55	55	123
11	66	89	199
12	78	144	322
13	91	233	521
14	105	377	843
15	120	610	1364
16	136	987	2207
17	153	1597	3571
18	171	2584	5778
19	190	4181	9349
20	210	6765	15127
21	231	10946	24476
22	253	17711	39603
23	276	28657	64079
24	300	46368	103682
25	325	75025	167761
26	351	121393	271443

27	378	196418	439204
28	406	317811	710647
29	435	514229	1149851
30	465	832040	1860498
31	496	1346269	3010349
32	528	2178309	4870847
33	561	3524578	7881196
34	595	5702887	12752043
35	630	9227465	20633239
36	666	14930352	33385282
37	703	24157817	54018521
38	741	39088169	87403803
39	779	63245986	141422324
40	819	102334155	228826127
41	860	165580141	370248451
42	902	267914296	599074578
43	945	433494437	969323029
44	989	701408733	1568397607
45	1034	1134903170	2537720636
46	1080	1836311903	4106118243

Table 39: First 46 Triangular Numbers, Fibonacci Numbers and Lucas numbers

Appendix B: *Math—Magic* with Paul Chika Emekwulu

B.1 Program Objectives

- Develops analytical and logical thinking skills.

- Presents the beauty, elegance, and excitement in number concepts.

- Entertains and stimulates interest through number tricks and investigatory lessons.

- Encourages pattern recognition.

- Actively involves and motivates students of all abilities.

- Encourages student-teacher, teacher-student, student-student communication.

- Supports the *National Council of Teachers of Mathematics* (NCTM) standards.

B.2 Program Description

Math—Magic is neither about magic nor numerology. Math—Magic is a program that uses creative and innovative teaching strategies to make mathematics exciting, interesting, and intriguing to high school students using paper and pencil. Most of these activities are embedded in guided discovery lessons that come in worksheet format.

Math—Magic is about motivation. It is about excitement. It is about mathematical reasoning. It is about pattern recognition. It is not about magicians' magic. It is not about numerology.

B.3 Past Engagements

Math—Magic has been presented to the following schools and associations:

- Oklahoma Council of Teachers of Mathematics

- Panhandle Mathematics & Science Conference

- Kansas Association of Teachers of Mathematics

- Booker T. Washington High School, Tulsa, OK

- Oklahoma City Community College,

Oklahoma City, OK

- Tecumseh High School, Tecumseh, OK

- Jenks Public Schools, Jenks, OK

- Newkirk High School, Newkirk, OK

- Norman High School, Norman, OK

- Oklahoma State University, Stillwater, OK

- Washington High School, Washington, OK

- Booker T. Washington High School, Tulsa, OK

- Oklahoma Education Association

- Liberated Arts Center, Oklahoma City, OK

- Oklahoma State University (Upward Bound), Oklahoma City, OK

- National Council of Teachers of Mathematics

- Board Members of Organization of Rural Oklahoma Schools

B.4 What People are Saying

"The looks on the students' faces were priceless and the 'Ah-Has' were abundant when the same equations and theories were introduced in new and interesting ways. We would recommend it to anyone who finds it difficult to grasp math concepts. 'Math—Magic' may be the key to unlock the mysteries of the math world."

Bennie Boykin
Upward Bound Director
Oklahoma State University
(Technical Branch), Oklahoma City)

"The Central Regional Conference Meeting of the National Council of Teachers of Mathematics in Topeka, KS will be called a resounding success because of people such as you who gave so generously of their time and effort. We are convinced that our conference was among the very best of regional conferences that NCTM has had."

Dr. Connie S. Schrock
Co-Program Chair
NCTM Central Regional Conference, Topeka, KS

National Council of Teachers of Mathematics
Central Regional Conference
St. Louis, Missouri

29–31 January, 1998

Dear Paul,

What do 400 excellent speakers, nearly 2400 participants, top-notch facilities, well-orchestrated arrangements, spring-like weather, and 'show-me'

Paul Emekwulu

hospitality equal? A memorable Saint Louis Central Regional Conference of the National Council of Teachers of Mathematics!

The accolades for the quality of the program and local arrangements are still arriving. Your presentation contributed to the praise the Program Committee continues to receive. We appreciate the time, thought, and preparation you gave to your part in the great success of our conference.

Best Wishes and many, many thanks,

Sincerely,

Carol A. Edwards

Program Chair

Organization of Rural Oklahoma Schools
Box 189
Foss, OK 73647

March 1, 1995

Paul Chika Emekwulu
Novelty Books
P.O.Box 2482
Norman, OK 73070

Dear Paul,
Thank you for your participation during the January meeting of the OROS board of directors.

The group was very pleased with your presentation. I will

be pleased to present members of the OROS board with copies of the "Program Request Form."

It is quite evident that you are able to capture the attention of your audience through "Magic with Numbers."

Sincerely,

Tom Butler

Executive Director, OROS

The University of Oklahoma
CENTER FOR THE STUDY OF SMALL RURAL SCHOOLS
COLLEGE OF CONTINUING EDUCATION

February 10, 1997

Mr. Paul Chika Emekwulu
Publisher
Novelty Books
P.O.Box 2482
Norman, OK 73070

Dear Mr. Emekwulu,

Congratulations! Your presentation entitled Math—Magic: Encouraging Mathematical Reasoning, Achieving Motivation and Excitement in the Classroom for the sixth annual National Conference on Creating the Quality School has been reviewed and accepted. The response to the call for presenters has been gratifying. The conference should prove to be exciting and valuable to attendees and presenters alike. The proposals were reviewed as quickly as possible by a panel to enable presenters ample notification to make travel plans. The conference begins Thursday, March 20 and concludes Saturday, March 22 at 10.30 a.m. (see enclosure).

A full registration brochure is included. Remember that registration is required for all presenters. Your pre-registration will help speed up the check-in process at the conference. To register by phone, call: 1-800-527-0772 ext. 2248. If you will not be able to attend, please notify us immediately so that we may allow someone else the opportunity to present. If you bring handouts, prepare for approximately 20-25 participants in your session. Please make your session as interactive as possible.

Please be sure your name, the name of your organization, and the address at the top of this letter are correct and as you would like them listed in the printed conference program. If there are changes, please write or fax them to us at your earliest convenience.

Please feel free to call us if you have any questions, concerns, or needs. This conference will address important issues, and we look forward to working with you. Again, contact us as needed at 800-937-4760 and ask for Cathie Parker.

See you in March!

Jan C. Simmons
Senior Program Development Specialist
Enclosures

Panhandle Mathematics and Science Conference
A Member of The Texas A & M University System
WTAMU Box 60208 Canyon, Texas 79016-0001 806-651-2626
Fax 806-651-2626

West Texas A & M UNIVERSITY
Division of Education

August 31, 2000

Paul Emekwulu
P.O.Box 2482
Norman, OK 73070

Dear Paul:

Re: Panhandle Mathematics and Science Conference

Thank you for your proposal for the above upcoming conference on Saturday, September 30th, 2000 titled, Math—Magic: Encouraging Mathematical Reasoning.

I am delighted to inform you that your proposal has been accepted, and that we are in the process of constructing the schedule for the conference at this time. The schedule, with times and room numbers, will be posted to our web site at www.wtamu.edu as soon as we have it completed. There is an icon on the bottom left-hand corner of the webpage that will lead you to the Panhandle Math/Science Conference site. You will be presenting your session 1 time(s) and we suggest that you prepare for 30 participants in each session.

Paul Emekwulu

Registration for the conference will begin at 8 am in the Jack B. Kelley Student Center on the WTAMU campus where there will be a speaker packet waiting for you. Lunch is provided and we hope that you will enjoy your day with us. If for some reason you cannot be with us, please be kind enough to let me know as soon as you can so that arrangements can be made to cancel your session.

If I can be of any further help, please do not hesitate to contact me by email at cpurkiss@mail.wtamu.edu or by phone at 806-651-2618. I look forward to seeing you on the 30[th]of September.

Sincerely,
Chris Purkiss
Chairperson

Appendix C: There Could be a Book in You

C.1 Program Objectives

In this seminar, the participants will:

- Gain inspiration and motivation that could translate into action.
- Realize that dreams and intuition could be sources of book ideas.
- Realize that only three things can stop a dream book.
- Realize that affirmations could be used to activate our creativity.

C.2 Program Description

How many of you have ever thought of writing a book?

Are you working on a book now, or have you ever submitted a manuscript to a publisher?

Do you think you have special knowledge or skills that you would like to share with others?

Are you a good story-teller?

Can you tell stories in a manner that holds people's attention?

Are you a teacher and have unusual but creative ways of presenting ordinary classroom concepts?

Are you currently doing seminars based on your experiences, and you don't have a book covering your topics?

Do you write articles for newspapers or magazines? If yes, have you ever thought of building a book out of these articles?

Do you have ideas for a book but don't know how to put them together?

Have you ever thought of collaborating with someone on a book project?

Do you have an idea of a particular cause that you would like to be remembered for?

If your answer to any of the above questions is *yes*, you need to write a book.

Emekwulu maintains that only three things can stop anyone from writing his or her dream book: lack of message, lack of faith in the message, and lack of faith in the messenger.

C.3 Past Engagements

- Norman Galaxy of Writers, Inc.

- Oklahoma City Writers, Inc.

- Canadian Valley Lions Club

- Metropolitan Library System, Mid-West City, OK

- Elk City Carnegie Friends of the Library, Elk City, OK

- Mid-Oklahoma Writers' Club

- Oklahoma Education Association

- Moore Association of Classroom Teachers, Moore, OK

About the Author

Paul Chika Emekwulu is an award-winning and international bestselling author. He is a co-author of an international bestselling book titled, *Unwavering Strength (Volume 2): Stories to Warm Your Heart & Soul.* When asked by the US Embassy officials in Lagos, Nigeria, why he petitioned for a non-immigrant entry visa into the United States, he said that he was coming to the United States to explore opportunity in publishing. Today, he is the author of six or more books including:

- *Mathematical Encounters for the Inquisitive Mind*

- *Mathematical Paradise: Getting to Know Triangular Numbers, Book One*

- *Mathematical Paradise: Getting to Know Triangular Numbers, Book Two*

- *Mathematical Explorations for Advanced Students*

- *Writing Down Your Dreams: Listen to Your Inner Voice and Change Your Life*

- *Divisibility Rules of Whole Numbers Made Simple* etc.

He has made invited presentations for schools, and organizations including:

- Oklahoma State University, Stillwater

- National Council of Teachers of Mathematics

- Oklahoma Education Association

- Moore Association of Classroom Teachers

- Kansas Association of Teachers of Mathematics

- Oklahoma Council of Teachers of Mathematics

- Oklahoma City Writers Inc.

- Norman Galaxy of Writers Inc.

- Black Liberated Arts Center

- 7 Hawks Publishing

- Mid-Oklahoma Writers Club

- Washington High School, Washington, OK

- Booker T. Washington High School, Tulsa

- Newkirk High School, Newkirk

- Norman High School, Norman, OK

- Michael Price School of Business, The University of Oklahoma

He is a member of Oklahoma Council of Teachers of Mathematics and Central Texas Council of Teachers of Mathematics.

He independently developed a mathematical formula connecting triangular numbers and numbers of the Fibonacci sequence. He has been a guest on several radio stations across the United States.

Useful Links

Meet the Co-Authors of *Unwavering Strength* Here
http://unwaveringstrength.com/co-authors/#PaulE

Social Media

Connect with me on Linkedin
https://www.linkedin.com/profile/view?id=65353103&trk=nav_responsive_tab_profile

Follow me on Twitter
https://twitter.com/pemekwulu

Befriend me on Facebook
https://www.facebook.com/authorpaulchika.emekwulu

Media Interviews

Listen to Paul's Interview with Cathryn Taylor of *Edge Magazine*
http://www.blogtalkradio.com/edgemagazine/2014/03/12/edge-inner-views-with-cathryn-taylor-and-paul-emekwulu

Listen to Paul's Interview with Anya Sophia Mann
http://anyasophiamann.com/quantum-alchemy/unwavering-strength/paul-emekwulu.php

www.ingramcontent.com/pod-product-compliance
Lightning Source LLC
Chambersburg PA
CBHW021430170526
45164CB00001B/174